Praise for Antidisestablishmentarianism

"...Great book if you want to know the truth about the Christian foundation our country was built ... "

" ... Great overview of the religious nature of the State Constitutions which were never barred from establishing religion by the U.S. Constitution."

"I enjoyed it!"

"Must Read!"

What Is Science?

(Serial Antidisestablishmentarianism Part Three)

by

Michael J. and Mary C. Findley

Findley Family Video Publications

What Is Science?

(Serial Antidisestablishmentarianism Part Three)

by Michael J. and Mary C. Findley

"Speaking the truth in love."

Table of Contents

Preface: Disestablishmentarianism

… When they knew God, they glorified him not as God, neither were thankful; but became vain in their imaginations, and their foolish heart was darkened. Professing themselves to be wise, they became fools…

Romans 1:21, 22

The most religious people on earth are those who claim not to have any religion. Dogmatic, intolerant, and bigoted, they refuse to allow anyone to so much as speak their opposition. Yet these same people demand political power and tax support. The mildest opposition, such as the mere mention of Intelligent Design (not God), has blacklisted tenured professors. Just two parents in a middle school in Texas made the national news by objecting to Gideon Bibles placed, without comment, on a table outside the school office.[1] Such people dishonestly claim that they are not religious and "religion" is a group of mythologies. The truth is that they are the ones promoting mythology. In every aspect of life they promote this mythology with unproven dogmatic assertions under the guise of "Science" vocabulary. After hijacking the word "Science," they use the courts to elevate their misuse of the term to an established religion.

Science is the study of the world around us, the use of the experimental method and the improvement of our lives through the application of technology. It is divided into various academic disciplines such as Chemistry, Physics, Mathematics and Biology. However, what the federal courts, the Academic community and the mainstream Western media mean by science is uniformitarianism. It is the cosmological foundation of the religion of Secular Humanism. “Since the fathers fell asleep, all things continue as they were from the beginning of the creation” (II Peter 3:4). This concise description of Uniformitarianism clearly shows that it is completely and entirely a religious belief in antiscientific myths.

Secular Humanists use words which have been in the English language for hundreds of years but give them “new” meanings. However, “there is no new thing under the sun” (Ecclesiastes 1:9, KJV). The words believe, faith and trust are all historic judicial terms and they also form the foundation of the true scientific method. What Secular Humanists promote as their version of the scientific method consists of preconceptions, presuppositions and assumptions. It is the opposite of an open mind.

A true open mind is founded in belief, faith and trust. The historic meaning of believe is to perceive or understand with the mind and then make an informed decision.[2] The most basic use of the word believe which the average American would understand is that of a juror in court. Which witness do you believe? Which piece of evidence is believable? A synonym would be the word credible. When we believe something or someone and then act on that belief, that is faith. The active part of belief is faith. The passive part of belief is trust.

Suppose your brother says that he will drive you to the doctor. If you believe him, then you understand what he says and you make a decision to get ready. If you get in the vehicle with him, that is faith. You act on your belief. When you sit in the vehicle as he drives, that is trust, a passive reliance on what you have proven true. You trust in his driving skills. You trust in the vehicle. You trust the roads, etc. Everything we do is a combination of belief, faith or trust. By restoring their historic definitions, belief, faith and trust re-emerge as the clear language of true experimental science. These terms were deliberately segregated from science to deceive people into believing Secular Humanism.

Liberals, Secular Humanists and materialists, however, use the word "belief" as a synonym for a philosophical position, just an opinion. Faith and trust to them are metaphysical words which mean different things to different people. And this is just the tip of an enormous iceberg. Secular Humanists have redefined hundreds of words to support their religion, such as sin, judgment and anthropology. A conversation with them can be very difficult since they use historical English words but mean something entirely different.

The traditional role of religion is to place priesthood as intermediary between God and man. The traditional role of an establishment of religion places the government in that intermediary role between God and man. In the Middle Ages the Roman Catholic Church put itself between man and God, as other religions have in the past. Johann Tetzel, a "professional pardoner," sold indulgences representing forgiveness for sins in Germany.

Indulgences were based on the "storehouse" of good works believed to exist because of the sacrifice of Christ and the good deeds and prayers of past saints. Tetzel was said to promise that, "As soon as a coin in the coffer rings, a soul from purgatory springs."[3]

Selling indulgences was the final act of many which brought on the Reformation. People wouldn't have bought them if they hadn't believed the Catholic Church alone could placate God on their behalf. Martin Luther convinced the princes of Germany that they did not need to send their money to Rome because they could go to God directly. Rome sent armies to collect the money. Even Modern Roman Catholics who do not believe that their church today claims to stand between them and God have to admit that the medieval Roman Catholic Church did.

The combined power of Church and State restricted personal worship, scientific study and access to historical truth. Today Secular Humanism has done the same by removing foundational truths from education. It excludes study and discovery that contradicts uniformitarianism. It rewrites history to undermine morality and freedom of expression.

The union between the medieval Romanist church and the state came to an end in two ways. In Southern Europe during the Renaissance, art, architecture, literature, and learning opened up to all men, not just those who were part of the church and state system. The Renaissance left the power intact, however. In Northern Europe, the Reformation abolished the need for a church like Rome through the great affirmations of the

Reformation: The Scriptures are the absolute authority; Justification is by faith alone apart from works; and every believer is his own priest with direct access to God. The Reformation made a special priesthood class unnecessary because men could pray directly to God and read His Word on their own.

The medieval Roman Catholic Church kept the Scriptures almost exclusively in Latin to prevent ordinary people from studying them, forcing people to come to the priest. The priest would not only tell them what the Scriptures said, but he also mingled that with the church's interpretation. In order for ordinary people who did not know Latin to read the Bible for themselves, the Scriptures had to be translated into the language of the ordinary people. Translation work by Reformers was essential to enable ordinary men to read the Scriptures for themselves, even though it was punishable by death under the Church-State system. The Renaissance and the Reformation worked together in the development of moveable type to make printing and distribution of translations of the Scriptures easier. Renaissance scholars revived interest in studying forgotten manuscripts and making translations into the vernacular. Erasmus's Greek New Testament provided a basis for more accurate translations of the Scriptures.

The Medieval Romanist Church-State system took away freedom by forcing man to rely on and accept its teachings. The Renaissance and the Reformation restored freedom by returning art, science, and all forms of learning to ordinary people. In particular the people were able to worship God as the

Scriptures taught, without Church-State control. Modern western culture, and American culture in particular, was founded on this religious freedom. American culture is more Christian than European cultures, but neither of these cultures can survive if the foundation of religious freedom is destroyed.

It is this Christian foundation of religious freedom which is the real target of Secular Humanists. These Secular Humanists have taken outrageous liberties in their unrelenting quest to replace religious freedom with their established religion of Secular Humanism, which they incorrectly call science or Natural Law. Their major tool is the US court system. Sympathetic US courts have consistently supported Secular Humanism by using every possible opportunity to replace the word religion with the ancient concept of Natural Law. However, since Natural Law has been used so many different ways, the courts had to standardize the term Natural Law. Their version of Natural Law goes back to Plato's *Republic*. Though Plato never used the phrase "natural law" in his *Republic,* translator Benjamin Jowett's notes state that, "Plato among the Greeks, like Bacon among the moderns, was the first who conceived a method of knowledge… "[4] Plato's *Republic* is at least the foundation of modern Natural Law, if not the detailed finished product. Together with Aristotle, Plato is supposed by secularists to have laid the foundation for learning and development of the Sciences. This is really is essence of Natural Law.

Jowett goes on to say that Plato provided for a means to spread his method of acquiring knowledge. "In the ideal State which is constructed by Socrates, the first care of the rulers is to be education."[4] Jowett makes it clear that Socrates meant to impart much more than mere academic knowledge, just as Natural Law means to teach more than mere Science. Socrates promoted "the conception of a higher State, in which 'no man calls anything his own,' and in which there is neither 'marrying nor giving in marriage,' and 'kings are philosophers' and 'philosophers are kings;' and there is another and higher education, intellectual as well as moral and religious, of science as well as of art, and not of youth only but of the whole of life."[4]

Many know that Plato in his *Republic* based his state on a philosopher/king. Few, however, are aware that he believed in communism and free love and that these two "natural" principles were to be foundational principles of the state.

Though the preceding condensation by Benjamin Jowett is an excellent job, as you can read for yourself, the actual words of Socrates, as quoted by Plato, are much longer and more difficult to understand. "None of them will have anything specially his or her own." "... Their legislator, having selected the men, will now select the women and give them to them [the legislator gives selected women to selected men]... they must live in common houses and meet at common meals ... they will be together ... And so they will be drawn by a necessity of their natures to have intercourse with each other..." "... Until philosophers are kings, or the kings and princes ... have the spirit and power of philosophy, and political greatness and wisdom meet in one ... cities will never have rest from their evils."[5]

The philosopher/king, according to Socrates, was to lay these foundational ideas through education. Though he did not use the phrase "establishment of religion," Plato clearly advocated an established religion. It was to be put in place by a philosopher/king through education based on a state where "no man calls anything his own" and where there is neither "marrying nor giving in marriage." Though this education would begin with children, it would continue throughout a person's entire life. This is the Natural Law which the US Court system has imposed.

The US needs to disestablish its Establishment of Religion and reestablish religious freedom. In the 1800's churches which tried to break away from the Church of England were called disestablishmentarians. The people who fought against the disestablishment of those churches within the Church of England in the 1800s were called Antidisestablishmentarians. Today, the mainstream media, liberal politicians, the academic community, the liberal courts and all others who file lawsuits, blacklist, fire, refuse to hire, tax, legislate against, libel, slander and do whatever is necessary to maintain their positions of privilege and power are modern Antidisestablishmentarians.

---

1 (No author) "Parents Fuming as Texas Schools Let Gideons Provide Bibles to Students," Tuesday, May 19, 2009, *Fox News.com*. "A spokeswoman for the school district said that a number of materials are made available to students this way, including

newspapers, camp brochures and tutoring pamphlets. College and military recruitment information is available all year long. The Gideon Bibles were made available for just one day. 'We have to handle this request in the same manner as other requests to distribute non-school literature – in a view-point neutral manner,' Shana Wortham, director of communications for the district, wrote in an e-mail to *FoxNews.com*.

2 Alexander Hamilton, in an 1802 letter to James Bayard. "I have carefully examined the evidences of the Christian religion, and if I was sitting as a juror upon its authenticity I would un-hesitatingly give my verdict in its favor. I can prove its truth as clearly as any proposition ever submitted to the mind of man."

3 Philip Schaff, *History of the Christian Church,* Volume 7, "The Reformation," Charles Scribner's Sons, 1910.

4 Plato, *The Republic* (c. 360 B.C.), translated by Benjamin Jowett over a period of 30 years until his death in 1893, completed posthumously by Lewis Campbell. (Introductory material (in double quotes) and paraphrases of Plato's ideas (in single quotes) were written by Jowett.)

5 Plato, *The Republic*, Book Five Dialogue excerpts among Socrates, Adeimantus, Glaucon and Thrasymachus have been placed in parentheses within Jowett's introductory material.

Introduction

Facts are stubborn things; and whatever may be our wishes, our inclinations, or the dictates of our passion, they cannot alter the state of facts and evidence.[1]

John Adams

Sometime in the early twentieth century, Secular Humanist indoctrination convinced almost everyone in the United States that "an establishment of religion" in the first phrase of the first amendment of the United States Constitution is vague and can mean just about anything. "The state of the facts and evidence," as John Adams so eloquently put it, is the exact opposite.

Section One of this work documents what the founders meant by the phrase "an establishment of religion." The Founding Fathers made as clear a statement as the English language permitted. The Constitution of the United States is founded on English law and to a lesser extent, various European laws, especially German and Dutch. In each of these countries, an Establishment of Religion was the collection of taxes to support education, welfare and public worship. The various governments appointed the teachers, welfare workers and pastors and expected these people to support the government in turn.

The original state constitutions not only permitted, but openly encouraged establishments of religion, especially in the areas of welfare and education. The foundation of the US Constitution is the fact that federal government was to have no control whatsoever in these areas. Their concept of a separation of Church and State was the exact opposite of what the courts have rammed down our throats for the past hundred years. The church should have the right to pray and teach without any federal intervention whatsoever. Judges should have the right to post any Scriptures they want. The courts should have no authority whatsoever to comment. Removing a state judge from office for posting the Ten Commandments is not merely an Establishment of Religion. It is the Inquisition.

Section Two documents the foundations of Secular Humanism and how it grew to become America's Establishment of Religion. The words "Secular Humanism" come from various groups in the 1950's. The phrase "Secular Humanist" is found in court documents to describe this set of beliefs. Secular Humanism is as old as civilization, but the primary foundation of twenty first century Secular Humanism is Plato's *Republic*. In America, Secular Humanism can be said to have originated with Thomas Paine. Secular Humanism has specific beliefs which are written down in various manifestos. Like Christianity, Islam and Judaism, Secular Humanism has many variations. Though Secular Humanists do not like the term, the most accurate words to describe these variants are "sects" or "denominations." Like Christians, Muslims and Jews, many Secular Humanist denominations do not get along with one another. Therefore, we have attempted to point out the beliefs which have the greatest agreement.

Section Three defines science, since Secular Humanists claim that science separates them from all other religions. Since true science is founded in the belief, faith and trust of the Bible, all of these words are defined carefully and in detail. In the Bible, belief, faith and trust are legal terms. Believe means to examine the evidence and come to a reasoned conclusion. Action taken on that belief is faith. Trust is the passive version of faith.

The Scientific Method is the biblical version of belief, faith and trust applied to the material world which God created for us. In the Bible, the Scientific Method recognizes that God is the creator, that we are required to be responsible managers of the material world God has given us, and that there is a final judgment after death which will include how well we managed the gifts God allowed us to use.

Our book concludes with Section Four, the results of having Secular Humanism as an Establishment of Religion. With the exception of America's founding documents and the ancient documents such as Plato, Plutarch and Genesis, hundreds of other quotes could easily be substituted for the quotes that appear here. There is nothing new or unique in this book. It is a combination of what used to be common knowledge in America before Secular Humanism took over and destroyed the education system and current events. If we were to start over today, we would pull different stories from the daily news. Though the individual stories would be different, the points would be the same. "There is nothing new under the sun" (Ecclesiastes 1:9). Or to state the same thing another way, the more things change, the more they stay the same.

America's Established Religion is Secular Humanism. This work is dedicated to exposing, defining and disestablishing it.

---

1 John Adams, "Argument in defence of the [English] soldiers in the Boston Massacre trial," December 1770.

2 "Alabama's Judicial Ethics Panel removed Chief Justice Roy Moore from office Thursday for defying a Federal judge's order to move a ten commandments monument from the State Supreme Court building." Friday, November 14, 2003. Posted 6:56 AM Eastern time. *CNN.com*

1. What Is Science to a Secular Humanist?

The sign of a natural law must be the universal respect in which it is held ... we would undoubtedly obey it universally ... Instead there is nothing in the world that is not subject to contradiction and dispute ... there is nothing that is strange and unnatural that is not approved in many countries[1]

Pierre Charron

All science is secular, and it did not originate in any supernatural faith.[2]

Charles Watts

"What is the relationship here of men to women?" Albert Einstein inquired during a 1921 visit to a Kibbutz in Israel. "*Herr Professor,* each man here has one woman." Answered the blushing female guide. Einstein quickly reassured her. "Don't be alarmed at my question - by 'relationship' we physicists understand something rather simple, namely how many men are there and how many women."[3]

The meanings of words can vary and so result in misunderstandings. Here "relationship" can mean a domestic arrangement as well as a ratio or comparison. The word "green" is an adjective which describes a color. This is commonly understood among English speaking people. Although green, like relationship, has other meanings such as envious, or inexperienced, or even ecologically friendly, few will misunderstand the sentence, "My shirt is green."

Without this simple, basic agreement on the meaning of words, language and any other form of communication, is impossible. Precise technical definitions of green in scientific fields, such as the exact wavelength in the electromagnetic spectrum that humans see as green, clarify and help avoid misunderstandings. This technical knowledge, however, is not necessary to understand what green means in ordinary conversation. The problem, not only with Secular Humanists, but also with liberalism in general, is that they use the same words as the rest of us, but change the meaning of the word. When a Secular Humanist uses the word science, he does not mean the same thing that a conservative does, and it is not a matter of mere misunderstanding.

To a Secular Humanist, science is defined by natural law. So we must understand what a Secular Humanist means by natural law before we can understand what he means when he defines the word science. To a Secular Humanist, natural law is similar to the idea of the laws of nature. That is, they are simply principles universally held to be true. This concept of natural law goes back at least to Plato and Aristotle. The problem with this view of natural law is that it is an erudite form of circular reasoning. Charles Watts defined truth as that which "the best knowledge endorses, the largest intellects accept, and the widest experience vouches for."[2] Put another way, something is true because it is universally held to be true. And that something is universally held to be true because they were able to convince, cajole, browbeat or force enough people to agree with them.

Secular Humanists also believe that truth is fluid and changeable. After they forced people to agree with them on their idea of truth, Watts went on to say, “Many so-called truths are liable to be corrected, modified, or superseded by more accurate power of judgment, or more perfect experience.”[2] Allowing for changeable truth removes all necessity for plan or purpose in nature. “Natural law” becomes whatever the “largest intellects” decide it is at the time. George Gaylord Simpson, paleontologist, said “Man is the result of a purposeless and natural process that did not have him in mind. He was not planned. He is ... a sort of animal... akin nearly or remotely to all of life and indeed to all that is material.”[4] This foundational dogma must also be imposed on unbelievers. But “unbelievers” (Theists) are stubbornly unwilling to be converted.

Julian Huxley demands that the children of these “unbelievers” must be indoctrinated with the dogma of their redefined natural law with no absolute truth. “It is essential for evolution to be-come the central core of any educational system, because it is evolution, in the broad sense, that links inorganic nature with life, and the stars with earth, and matter with mind, and animals with man.”[5] He insisted on indoctrination with evolution, wanted it “central” to education, because he demanded belief in progression from lower to higher, primitive to advanced. This is the dogmatic opposite of Christianity, which teaches that man was created in the image of God with the constant struggle to conserve the original creation or to degrade from it. Julian Huxley wanted to "prove" that animal nature and human nature were not distinct, that there was no soul or spiritual nature breathed in by God. "Human history is a continuation of biological evolution in a different form."[5] This is as purely a religious statement as it is possible to make. It is simply an expression of Secular Humanism with the power of an Established Religion. This is what a Secular Humanist means when he uses the word "science."

These same people who say that reason is the supreme "natural law," however, justify the most depraved behavior when passions overwhelm reason. To a Secular Humanist, a teenager has raging hormones with uncontrollable sexual drives. When reasonable people point out that this view of science strips away morality, Richard Dawkins explains: "Scientific and technological progress themselves are value-neutral. They are just very good at doing what they do. If you want to do selfish, greedy, intolerant and violent things, scientific technology will provide you with by far the most efficient way of doing so."[6] The presumption is that "real" scientists would only use it for good.

This, to Secular Humanists, is one of many "natural laws" which provide their foundation for a "Brave New World" of science. "But if you want to do good, to solve the world's problems, to progress in the best value-laden sense, once again, there is no better means to those ends than the scientific way."[6] This arrogance is the opposite of the Natural Law of America's founding fathers.

There is, however, one aspect of Natural Law with universal agreement. Natural Law should explain universal truths of nature. "To explain all nature is too difficult a task for any one man or even for any one age,"[7] said Isaac Newton, a recognized proponent of Natural Law. "'Tis much better to do a little with certainty, & leave the rest for others that come after you, than to explain all things by conjecture without making sure of anything."[7] It is sad that Secular Humanists usually ignore this point of agreement and emphasize their doctrine of natural laws relying heavily on conjecture. "The sign of a natural law must be the universal respect in which it is held ... we would undoubtedly obey it universally ..." Clearly this is not the case. "Instead there is nothing in the world that is not subject to contradiction and dispute ... there is nothing that is strange and unnatural that is not approved in many countries ..."[1] Charron is making the logical transition from the laws of nature to natural law as it applies to man. Julian Huxley envisioned a wholly "scientific" reason-based natural law. That vision invariably produces the result Charron described.

What a Secular Humanist means by "scientific" is materialism. It is the foundation of an all-consuming religion that crowds God out of every aspect of everyone's life. "Science is the sole available means of temporal help,"[2] said Charles Watts, president of the National Secular Society in Great Britain. "Science" as defined by the secularist replaces God. They believe that all other religions are either opposed to science or at least operate outside of the realm of science. According to Watts, what Secular Humanists mean by the word science is, “All science is secular, and it did not originate in any supernatural faith.”[2]

However, they measure their own faith with a different standard. “In tracking the emergence of the eukaryotic cell, one enters a kind of wonderland where scientific pursuit leads almost to fantasy. Cell and molecular biologists must construct cellular worlds in their own imaginations.... Imagination, to some degree, is essential for grasping the key events in cellular history”[8] Secularists refer to themselves as “scientists” and call biblical belief “fantasy” and “imagination.” Yet they use the same terms to describe what they consider legitimate “research.” Unless we define the same words the same way, we cannot have a conversation.

Secular Humanists, however, demand freedom to speak about their definition of science in terms curiously similar to Divine Revelation while ridiculing religion. Richard Dawkins insists that it must not be thought that there are some things the dogma of secularism cannot explain. “Just because science so far has failed to explain something, such as consciousness, to say it follows that the facile, pathetic explanations which religion has produced somehow by default must win the argument is really quite ridiculous.”[9]

At some point secularists assure us that we will understand all the truth and discover all there is to know about nature and its workings, but for now there is knowledge that is hidden from us and only “free inquiry” will permit it to be discovered. “An unfolding, opening-out, or unwinding; a disclosure of something which was not previously known,” says Charles Watts of evolution, “but which existed before in a more condensed or hidden form. According to this theory, there is no new existence called into being, but a making conspicuous to our eyes that which was previously concealed."[2] This is not a scientific theory. It is a faith more dogmatic than any other religion.

According to Charles Watts, it has replaced the Bible for things humans do not understand. This dogmatic faith replaces the Divine Revelation in the Scriptures which God uses to clarify things we did not know or understand before, but which always existed. According to Richard Dawkins, the revelations of evolutionary knowledge will be trumpeted through its crowning achievement, man. "We humans are an extremely important manifestation of the replication bomb, because it is through us – through our brains, our symbolic culture and our technology – that the explosion may proceed to the next stage and reverberate through deep space."[10]

This is what Secular Humanists mean by scientific "discovery." Secularist scientific "discovery" depends heavily on the ability to speculate about what might be learned, to presuppose and theorize until the theories can be declared to be facts without proof. "Scientific beliefs are supported by evidence, and they get results. Myths and faiths are not and do not."[11] Richard Dawkins dogmatically states that his scientific "beliefs" are superior and he dismisses what he calls religious faith. Secularist revelation consists of using studies to announce that speculation has become a theory and eventually that theory is fact.

By contrast, America's Founding Fathers based their documents such as the Declaration of Independence, Constitution of the United States and the various state Constitutions, on a very different kind of Natural Law. They defined Natural Law as rules God had set up at Creation. Natural Law operated predictably and could be discovered and understood using reason. America's Founding Fathers believed God must guide reason. They also believed that reason is practiced in a universe created and sustained by God. This is very different from the "reason" of Secular Humanism.

"Reason has built the modern world,"[12] Richard Dawkins said. Reason properly applied could lead men to make good laws or good science. "It is a precious but also a fragile thing, which can be corroded by apparently harmless irrationality."[12] The Founding Fathers apparently were guilty of "harmless irrationality." "We must favor verifiable evidence over private feeling"[12] Almost all of America's Founding Fathers held religious "feelings," public or private. "Otherwise we leave ourselves vulnerable to those who would obscure the truth."[12]

This is dishonesty at its worst.

"We have gone back to ancient history for models of government," said Benjamin Franklin at the Constitutional Convention, "and examined the different forms of those Republics."[13] The Founding Fathers could also prove that they had arrived at their conclusions about government by the same reasoning processes that produced scientific knowledge. They determined that other republics, "having been formed with the seeds of their own dissolution now no longer exist."[13] Their reason told them through a scientific process of investigation that other governmental forms were flawed. "And we have viewed Modern States all round Europe, but find none of their Constitutions suitable to our circumstances."[13]

Benjamin Franklin, possibly the most rational and scientific-minded, as well as the least likely among all the Founding Fathers to suffer the name of Christian, still insisted that human reason must be guided by God in order for Natural Law to operate as it should. "How has it happened, Sir, that we have not hitherto once thought of humbly applying to the Father of lights to illuminate our understandings?"[13]

Franklin was a man of science, experimenter, discoverer, inventor, and yet he acknowledged that "... God governs in the affairs of men. ...Is it probable that an empire can rise without his aid? 'Except the Lord build ... they labor in vain...'"[13] Franklin even feels free to quote and refer to Scripture in a government meeting. "...Without his concurring aid we shall succeed in this political building no better than the Builders of Babel: ...Mankind may hereafter ... despair of establishing Governments by Human Wisdom, and leave it to chance, war, and conquest."[13] Note that Franklin believes in the power of human wisdom, but insists it needs God's guidance to understand how Natural Law should work.

Even before the Constitution came to be formed, with the example of the "scientific" reliability of these natural laws before him, Samuel Adams could see just as clearly that there were natural rights of man. These were rules the Creator intended man to live by. As Franklin said, they were based on the observations and experimentations of men through the centuries to discover what was and was not true and right government.

Adams wrote that the natural rights of the colonists were life, liberty and property. "Among the natural rights of the Colonists are these: First, a right to life; Secondly, to liberty; Thirdly, to property; together with the right to support and defend them in the best manner they can."[14] He also believed that everyone naturally retained all rights that they did not voluntarily relinquish. "These are evident branches of, rather than deductions from, the duty of self-preservation, commonly called the first law of nature."[14] Since slavery existed in the colonies, Thomas Jefferson had to change the natural right of property to the pursuit of happiness, another acknowledged principle of Natural Law. Man had the opportunity to pursue happiness. Because he had a fallen nature, he did not have the guarantee that he would unfailingly obtain it.

"The greatest happiness of the greatest number,"[15] is the mantra of secularism. To secularists happiness depends on all men as a group, whereas the Founding Fathers clearly understood the pursuit of happiness to be an individual freedom. "All positive and civil laws should conform, as far as possible, to the law of natural reason and equity."[14] Others believed that all mankind had other natural rights such as the right to exercise freedom of religion, to enter into social contracts, to be free from tyranny and the freedom to use their reason. "As neither reason requires nor religion permits the contrary, every man living in or out of a state of civil society has a right peaceably and quietly to worship God according to the dictates of his conscience."[14] These men wrote this view of natural law into these documents, the laws they passed, their personal correspondence and judicial decisions. This is not, however, how Secular Humanists define natural law.

Science to a Secular Humanist is the systematic study of the material universe and nothing more. "The feeling of awed wonder that science can give us is one of the highest experiences of which the human psyche is capable."[16] Yet Richard Dawkins, like Carl Sagan, describes it as if it were a religious experience. "It is a deep aesthetic passion to rank with the finest that music and poetry can deliver."[16] It has to be as fulfilling as a religion for the secularist, because he is quick to point out that there is no heavenly bliss to look forward to. "It is truly one of the things that make life worth living and it does so, if anything, more effectively if it convinces us that the time we have for living is quite finite."[16] He founds this conclusion in the religious belief that the material universe is all that is, was or ever will be. Though he claims a completely fact-based scientific viewpoint, the secularist is also fond of claiming that he can actually see and measure the universe as a whole. Carl Sagan did it, and many other cosmologists have confidently taken us on journeys outside the known boundaries of our universe.

Science textbooks and university-funded scientific websites promote this mythology, as technical writer Jonathan David Carson points out. "The textbook Astronomy Today by Eric Chaisson and Steve McMillan depicts the universe, which it says has a diameter of 1028 mega parsecs, as a 2 1/2inch sphere, implying a point of view 6 x 1028 mega parsecs outside the universe. The University of Michigan's Windows to the Universe Website depicts the universe as a 1/2-inch sphere, implying a point of view 300 billion light-years outside the universe."[17] Carson brings up multiple examples of this supposed fact-based point of view, which he rightly concludes puts man into an arrogant, godlike state of observing from outside the entire universe. Again "science" turns man into God, opposing the fact that man is and that man's view of the universe is very finite.

Like any religion which enthrones man in God's place, there is a desperate and irrational need to attack true religion. "Faith is the great cop-out, the great excuse to evade the need to think and evaluate evidence,"[18] says Richard Dawkins. In the Bible, in the founding documents of US history and in the US court system prior to the liberal takeover, belief was (and still is in reality) a legal term. Belief is the decision of a juror based on evidence. Faith is the action one takes based on belief based on tested evidence. The modern Secular Humanist twists the word "faith" to mean the opposite of its historical definition. "Faith is belief in spite of, even perhaps because of, the lack of evidence."[19] This is the "blind leap of faith" of Karl Barth and neo-orthodoxy, not the historic meaning of faith found in the Bible and US history.

The faith of the secularist, which is truly "in spite of, even perhaps because of, the lack of evidence," has a religious belief that the material universe is all that is, was or ever will be. The material universe is the ultimate reality. "Who is more humble?" asked Carl Sagan, "The scientist who looks at the universe with an open mind and accepts whatever the universe has to teach us, or somebody who says everything in this book [the Bible] must be considered the literal truth and never mind the fallibility of all the human beings involved?"[20] Sagan is pretending humility while arrogantly dismissing the possibility that God might have actually written down His words out of love for his creation.

"Anything you don't understand, Mr. Rankin, you attribute to God,"[21] says a character in Carl Sagan's novel *Contact*. A Secular Humanist believes ultimate truth must be analyzed, measured and described with material tools and the results of all experiments must be material. "God for you is where you sweep away all the mysteries of the world, all the challenges to our intelligence. You simply turn your mind off and say God did it."[21] In the same work Sagan pretends to be skeptical but open-minded when he has a character say, "The major religions on the Earth contradict each other left and right. ... You must care about the truth, right?"[21]

Truth always contradicts error. The way to find truth is not to arrogantly proclaim yourself to be a skeptic, but to be honest. "Well, the way to winnow through all the differing contentions is to be skeptical."[21] He has already dismissed religions of all kinds because they "contradict each other" and he claims that they could all be wrong yet he claims objectivity. "I'm not any more skeptical about your religious beliefs than I am about every new scientific idea I hear about."[21] Yet conclusions which do not have material explanations are rejected out of hand. "But in my line of work, they're called hypotheses, not inspiration and not revelation."[21] These hypotheses are supposed to be based on data. To a Secular Humanist, however, data which does not agree with a priori assumptions is sometimes altered but usually just ignored. Honesty, which might involve some skepticism at times, is the real way to arrive at truth.

Science to a Secular Humanist is the infallible dogma used to put down and hopefully silence opposition while maintaining political power. This political power is used to maintain educational funding, provide audiences for religious propaganda through museums, television and radio programs, organizations such as the National Geographic society, wording on plaques at national parks, zoos and aquariums, paying editors of newspapers, magazines and web blogs and putting signs up by the millions which state that their religion is fact and all other religions are false.

"Most people, I believe, think that you need a God to explain the existence of the world, and especially the existence of life. They are wrong, but our education system is such that many people don't know it."[22] Either Richard Dawkins actually believes that the education system is not completely secular, or he is just impatient that it is not overcoming belief fast enough.

Organizations such as National Geographic support Secular Humanism in every magazine issue, nearly every article or photograph. "A prehistoric snake is poised to make a meal of a newly hatched dinosaur in this life-size reconstruction of 67-million-year-old fossils unearthed in India."[23] In 2010, Grand Canyon National Park is set to unveil an exhibit funded by the National Science Foundation in 2006.

> *"The Trail of Time, the world's largest interpretive geoscience exhibition ... a horizontal timeline that can be walked along 2 kilometers of the existing paved South Rim Trail; it is marked with inset 5-cm circular bronze markers at one-meter intervals, and 10-cm numerically labeled circular bronze medallions every 10 meters (Figure 2). Each meter along the Trail of Time represents one million years of elapsed time: 1 meter = 1 million years. ... [Another] segment is intended to help Grand Canyon visitors adjust their temporal frames of reference ... through historic and archaeological time scales (centuries to millennia) to deep time (millions of years)."*[24]

This "timeline" even contradicts the most infallible standard Secular Humanists, radiometric dating. According to radiometric dating, the top layer of the Grand Canyon is the oldest layer.[25] Also according to radiometric dating, none of the layers progress to the next layer in the date order presented by "The Trail of Time." Even pointing out this factual error is considered an "attack."

Like priests of ancient Greek temples, Secular Humanists try to shield their dogma from "attacks." "Attackers" include proponents of Intelligent Design such as the Discovery Institute, which describes itself on its website as a non-profit, non-partisan public policy center for national and international affairs. (See Section Two Discovery Institute Appendix for more information) Jason Rosenhouse, PhD in Mathematics and educational observer, quotes from an article in *First Things* by University of Delaware physics professor Stephen Barr.

> *[Barr] Lays into the ID Movement. Here's the first paragraph: "It is time to take stock: What has the intelligent design movement achieved? As science, nothing. The goal of science is to increase our understanding of the natural world, and there is not a single phenomenon that we understand better today or are likely to understand better in the future through the efforts of ID theorists. If we are to look for ID achievements, then, it must be in the realm of natural theology. And there, I think, the movement must be judged not only a failure, but a debacle."*[25]

This is the same technique used by Adolph Hitler in his 1925 *Mein Kampf* ("My Struggle") and clarified in 1928 by the Nazi propaganda minister Joseph Goebbels. In America, we call this technique "the big lie." Barr's statement is propaganda from the beginning, with no hint of fact.

The authors drove along the winding roads through the Bighorn Mountains in Wyoming. Signs everywhere proclaim the uniformitarian geologic eras. A photo description on the Wisconsin University website reads as follows.

> *Big Horn Mountains, WY- Piney Creek Thrust. View to northwest along strike of*

eastward-thrust Paleozoic formations. Precambrian crystalline rocks lie at the left. Cambrian sediments crop out in the grassy slopes. The Bighorn dolomite (Ordovician) and Madison limestone (Mississippian) form the ridge in the central part of the photo. Tertiary erosion debris--i.e. Moncrief member of Wasatch fm--is at the right.[26]

Richard Dawkins commented after the 9/11 terrorist attacks that he wanted the statement below read at his funeral. Dawkins stands before the world as a man of science, supposedly a lofty example of what education, "free inquiry" and reasoned study can produce. Yet he makes the most ignorant, narrow-minded statements about religion imaginable. He considers all religions to be the same, all to be nonsense. But now it is equally clear to him that all religions are dangerous. The man of science demands that everyone stop respecting religion and join him, not really in just criticizing it, but in stamping it out. (Please excuse the inclusion of the profanity in the quote but it is both a demonstration of Dawkins' religious fervor and his determination to be crude and insulting to believers.)

"Many of us saw religion as harmless nonsense. Beliefs might lack all supporting evidence but, we thought, if people needed a crutch for consolation, where's the harm? September 11th changed all that. Revealed faith is not harmless nonsense, it can be lethally dangerous nonsense. Dangerous because it gives people unshakeable confidence in their own righteousness.

> *Dangerous because it gives them false courage to kill themselves, which automatically removes normal barriers to killing others. Dangerous because it teaches enmity to others labelled only by a difference of inherited tradition. And dangerous because we have all bought into a weird respect, which uniquely protects religion from normal criticism. Let's now stop being so damned respectful!"*[27]

---

1 Pierre Charron, *De la sagesse* (Of Wisdom, In Three Parts), French version, 1601, Translated by Samson Lennard, Eliot's Court Press for Edward Blount and Will, Aspley, London, c.1615.

2 Charles Watts, "The Secularist's Catechism," complied in an undated book published by Watts & Co. entitled: *Pamphlets by Charles Watts,* Vol. I, originally written in 1896.

3 Adapted from Einstein, *A Life,* Denis Brian, John Wiley and Sons, New York, 1996.

4 George Gaylord Simpson, *The Meaning of Evolution,* revised edition, New Haven: Yale University Press, 1967.

5 Julian Huxley, "At Random," television preview Nov. 21, 1959.

6 Richard Dawkins, *The Root of All Evil,* January 2006 television documentary.

7 Unpublished notes for the *Preface to Opticks* (1704) quoted in *Never at Rest: A Biography of Isaac Newton* (1983) by Richard S. Westfall.

8 B.D. Dyer & R.A. Obar, *Tracing the History of Eukaryotic Cells,* Columbia University Press, 1994, pp. 2,3.

9 Richard Dawkins, quoted in "The Evolutionary Future of Man," *The Economist,* 1993-09-11, vol. 328, p. 87.

10 Richard Dawkins, quoted in "The Flying Spaghetti Monster," by Steve Paulson, *Salon.com,* October 13, 2006.

11 Dawkins, *River Out of Eden,* Basic Books, the Perseus Book Group, New York, NY, 1995.

12 Dawkins, "Slaves to Superstition," *The Enemies of Reason,* [1.01], timecode 00:46:47ff, aired 13 August 2007.

13 Benjamin Franklin, Speech to the Constitutional Convention (28 June 1787).

14 Samuel Adams, *The Report of the Committee of Correspondence to the Boston Town Meeting,* Nov. 20, 1772.

15 Jeremy Bentham, *The Works of Jeremy Bentham,* vol. 4, 1843, 11 vols, Edinburgh: William Tait, 1838-1843.

16 Richard Dawkins, *Unweaving the Rainbow: Science, Delusion and the Appetite for Wonder,* Houghton-Mifflin, New York, NY, 1998.

17 Jonathan David Carson, "Science's Sins of the Eyes," *New Oxford Review,* November 2001.

18 Richard Dawkins, from a speech at the Edinburgh International Science Festival, April 15, 1992.

19 Dawkins, *The Richard Dimbleby Lecture:* "Science, Delusion and the Appetite for Wonder," *BBC1 Television* November 12, 1996.

20 Carl Sagan, in an interview with *Charlie Rose*, late-night PBS talk show host, 1996.

21 Sagan, *Contact,* Pocket Books, Simon and Schuster, New York, NY, 1985.

22 Richard Dawkins, "From Tail to Tale on the Path of Pilgrims In Life," *The Scotsman,* April 9, 2005.

23 *National Geographic*, March 1, 2010, photo caption.

24 From the Arizona State University website, an undated article by Steven Semken, Associate Professor of Geoscience Education and Geological Sciences, School of Earth and Space Exploration at Arizona State University.

25 Duane T. Gish, "A Decade of Creationist Research" (Part I), *Creation Research Society Quarterly* 12 (1):34-46 June, 1975.

26 Jason Rosenhouse, *EvolutionBlog,* posted March 5, 2010.

27 The *Wisconsin University website,* overview of the Laramide/ Yellowstone mountain ranges with aerial maps designating geologic ages.

28 Richard Dawkins, in a speech following the 9/11/2001 Islamic-led terrorist attacks on targets in the United States.

2. What Is True Science?

Pilate saith unto him [Jesus],What is truth?

John 18:38 KJV

Secularists like Charles Watts and Richard Dawkins affirm a single belief, though they vary in their methods of stating it. They believe that there is no evidence for the truth of any "revealed faith" and that only secularist science should be permitted to exist. People who call a religion "revealed" mean that somehow these people were told what to believe by a non-material means and that there is no outside evidence that the written record, their holy scriptures, are true or authoritative. Sam Harris, author of The End of Faith, said, "We have names for people who have many beliefs for which there is no rational justification. When their beliefs are extremely common we call them 'religious'; otherwise, they are likely to be called 'mad', 'psychotic' or 'delusional'..."[1] That's why they call it revealed faith, because secularists have changed the meaning of the word faith into blindly accepting something without evidence.

Every ancient polytheistic belief system has stories of outrageous behavior by its gods. Buddhism and similar "wisdom" religions have no gods per se but they rely upon "enlightenment" received in an otherworldly state of solitary meditation. Islam was delivered to Mohammed in the form of the Koran. The Latter Day Saints received their revelatory scriptures on golden plates from the angel Moroni. Andy Rooney once said about "revealed faiths" that "Those to whom his word was revealed were always alone in some remote place, like Moses. There wasn't usually anyone else around when Mohammed got the word, either. Mormon Joseph Smith and Christian Scientist, Mary Baker Eddy, had exclusive audiences with God. We have to trust them as reporters - and you know how reporters are. They'll do anything for a story."[2]

Secularists delight in saying that all religions are equally unreliable, their revelations received in secret. But in the case of Moses, he received a relatively brief commission alone in the desert at the burning bush. The children of Israel did not say they were forced to swallow a tall tale of Moses' secret revelation. They said, "The LORD our God hath shown us his glory and his greatness, and we have heard his voice out of the midst of the fire." Most of the written laws of God were delivered in full view of the entire nation of Israel on Mt. Sinai. "We have seen this day that God doth talk with man, and he liveth. Now therefore why should we die?" (Deuteronomy 5:24, 25, KJV) The people, 603,550 fighting men plus women, children and the aged, (Numbers 1:46) were so convinced that God was really there in the thunder and lightning, and that the laws came directly from Him, that they begged Moses to go up on the mountain as their representative, convinced that as frail sinners they could die in the holy presence of God. Every Israelite could see the cloud and pillar of fire in the middle of the camp every day. All of them saw the plagues in Egypt, participated in the Red Sea crossing, watched Pharaoh's army drown in the Red Sea, heard God speak to Moses at the tabernacle and saw the destruction of those who rebelled. None of these events took place in secret locations or for a chosen few.

In many cases prophets in the rest of the Old Testament spoke openly to the people as God gave them a revelation. "And Elijah said unto all the people, Come near unto me. And all the people came near unto him. And he repaired the altar of the LORD that was broken down. And Elijah took twelve stones, according to the number of the tribes of the sons of Jacob, unto whom the word of the LORD came." (I Kings 18:30, 31, KJV) John the Baptist was in the desert but crowds of people flocked to him. He didn't collect them after he got some secret revelation.

> *"Then went out to him Jerusalem, and all Judaea, and all the region round about Jordan, And were baptized of him in Jordan, confessing their sins." (Matthew 3:5,6, KJV) John the Baptist gives testimony that the miraculous circumstances at Jesus' baptism occurred in full view of all the people present with him at the event. "And John bare record, saying,' I saw the Spirit descending from heaven like a dove, and it abode upon him.' ...'Upon whom thou shalt see the Spirit descending, and remaining on him, the same is he which baptizeth with the Holy*

> *Ghost.' And I saw, and bare record that this is the Son of God."*
>
> (John 1:32-34, KJV).

The teachings and attesting miracles (attesting means proof that He was who He said He was) of Jesus Christ in the New Testament were usually in open places before hundreds or thousands of people. "And Jesus, when he came out, saw much people, and was moved with compassion toward them, because they were as sheep not having a shepherd: and he began to teach them many things. ... And they that did eat of the loaves were about five thousand men." (Mark 6:34, 42-44 KJV)

This lie (teaching in secret) was one of the false charges leveled against Jesus at his mockery of a trial. Though Jesus said very little at these so-called trials, he did answer this one charge. "Jesus answered him, 'I spake openly to the world; I ever taught in the synagogue, and in the temple, whither the Jews always resort; and in secret have I said nothing. Why askest thou me? ask them which heard me, what I have said unto them: behold, they know what I said.'" (John 18:20,21, KJV)

Another major argument used is that religions contradict each other and they cannot all be right. *Amazon.com* has a question/comment site where one contributor asked, "Many religious people today say that all traditional religions are equal: all are 'true' to those who believe. Truth is in the eye of the beholder. Does this mean they are all equally false? Or what?"[3] Truth is not subjective. It is exclusive. There can be only one true belief. You can choose to be right or wrong. Over the centuries millions of people have chosen to base their belief on the Bible, specifically on Christianity as the culmination of the Old and New Testament Scriptures and the only reliable and true faith. Why have millions of people trusted in and died for Jesus the Messiah? Christianity begins and ends with Jesus the Messiah and the authority of the Scriptures.

There are two and only two sources for learning about Jesus: the Bible and other people outside of the Bible who wrote about Jesus. Nothing is more important than honesty in this investigation. Nothing is more important than correctly evaluating the evidence. The *Titanic* sank with half empty lifeboats because people could not be convinced that the great ship was sinking. Witness reports vary greatly on events at the time of the sinking but one source affirms that those passengers who survived acted quickly and did not delay while waiting to be certain they were in peril.

> *...The crew worked quite efficiently, taking a total of 80 minutes to lower all 16 lifeboats. ... the high death toll that related to the lifeboats was the reluctance of the passengers to board them. They were on a ship deemed to be 'unsinkable.' Because of this, some lifeboats were launched with far less than capacity, the most notable being Lifeboat #1, with a capacity of 40, launched with only 12 people aboard. Included in the first launched were lifeboats 6, 7, and 8, each of which were equipped to hold 65 but evacuated the ship with only 28 on board each boat.*[4]

The lifeboats had to move away from the *Titanic* in the final minutes as she slipped beneath the surface.

Crewmembers manning the boats believed they would have been pulled down with it. Too late, the people remaining onboard the *Titanic* realized that the great ship was sinking rapidly. The lifeboats had already moved out of reach.

More certainly than the *Titanic's* demise, each and every person reading or listening to this knows that he is going to die. Each of us has the responsibility of correctly evaluating the evidence. At the trial of Jesus, the Roman procurator Pontius Pilate asked the simple question, "What is Truth?" Courts have complex rules to keep out tainted evidence and guarantee that only the relevant facts are presented. Though most people have more than the few minutes which the passengers of the *Titanic* had or the few weeks of a court case, our lives are still, as James said, "like a vapour." (James 4:14) Evaluating the evidence for Jesus Christ is more important than who we marry, how much education we have, what we choose for a career, where we work, where we live, what we wear or eat or who we talk to.

"Now there was about this time Jesus, a wise man, if it be lawful to call him a man; for he was a doer of wonderful works, a teacher of such men as receive the truth with pleasure." Josephus is the only first century witness to Jesus the Messiah outside of the Bible. "He drew over to him both many of the Jews and many of the Gentiles. He was [the] Christ." He was a small child at the time of Jesus' ministry but was able to interview people who directly knew Jesus. "And when Pilate, at the suggestion of the principal men among us, had condemned him to the cross, those that loved him at the first did not forsake him; for he appeared to them alive again the third day, as the divine prophets had foretold these and ten thousand other wonderful things concerning him." Some dispute whether Josephus actually wrote this, or was even a believer himself, but it is a near contemporary, corroborating account. "And the tribe of Christians, so named for him, are not extinct at this day."[5]

Witnesses among the early church fathers such as Clement or Ignatius go back to the early second century, but not to the time of Jesus. "The four earliest Christian writers who report the date of Jesus' birth are Irenaeus (late second century), Clement of Alexandria (about A.D. 200), Tertullian (early third century), and Africanus (early third century)."[6] As a group these valid witnesses are known as secondary or circumstantial witnesses. The only primary witness to Jesus as the Messiah is the Bible itself. The major reason for this is the zeal the Romans had for destroying any reference to Jesus of Nazareth. "This was the nineteenth year of the reign of Diocletian in Dystrus (which the Romans call March), [303 AD] ... royal edicts were published everywhere, commanding that the churches should be razed to the ground, the Scriptures destroyed by fire..."[7] They burned any Christian writings and put Christians to death by the thousands for possessing them. These persecutions were so intense and so efficient that believers had to decide which documents they were willing to die for.

By contrast, some of Plato's writings do not have a single copy written within a thousand years of his death. "Oxford's most important manuscript of classical philosophy is the *Clarke Plato* (MS. E. D.

Clarke 39), the oldest surviving manuscript for about half of Plato's dialogues, which was acquired by the University in 1809: it was written in Constantinople in A.D. 895."[8] A single, carefully preserved, fragmentary manuscript labeled "Plato, *Phaedo* 3rd c. BC Brit. Mus. Pap. 488" exists as a papyrus in the British museum. (Plato lived c.427-c.347 BC).

Most ancient documents have less than one dozen early manuscripts, many only a single copy. Yet few seriously question their authenticity. "The Rylands Papyrus (P52)" is a fragment of the New Testament. "This papyrus was found in Egypt, and has been dated at about 125 A.D."[9] *Codex Vaticanus* and *Codex Sinaticus* date from the 4th century and earlier. The New Testament has more than four thousand copies and fragments from the first through the fourth centuries AD. These documents were preserved at the cost of their owners' lives. "Tertullian stated that by 150 A.D., the Church in Rome had compiled a list of the New Testament books matching our list of today."[9] No other ancient writing has as much documentation. And still, more people question the Bible's authenticity than any other ancient document. If the Bible is not an authentic document, written when, where and how it claims to be, then the same rules of evidence must apply to all other ancient documents and no other ancient document can be authentic either.

It is possible, however, for the Bible to be an authentic document, written in the early to middle of the first century by the people whose names are attached to each book, but the information they wrote might still be in error. The Bible, however, claims to be the Word of God. The Scriptures cannot be broken (John 10:35). Heaven and Earth shall pass away, but my words shall not pass away (Matthew 24:35). All Scripture is given by inspiration of God, and is profitable for doctrine, for reproof, for correction, for instruction in righteousness (II Timothy 3:16). It is either everything it claims to be or it is a lie. There is no middle ground. If Jesus was not God, who died as a sacrificial atonement for our sins and rose again the third day, then He was a liar, a fraud and a charlatan. The Bible is not just a good book. It is either the inspired Word of God or a lie. Jesus is more than just a good example. He is completely and entirely what the Bible claims or the Bible is not authoritative. The Bible goes even further and condemns all other books as false.

> *What thing soever I command you, observe to do it: thou shalt not add thereto, nor diminish from it (Deuteronomy 12:32). Every word of God is pure: he is a shield unto them that put their trust in him. Add thou not unto his words, lest he reprove thee, and thou be found a liar (Proverbs 30:5,6). For I testify unto every man that heareth the words of the prophecy of this book, If any man shall add unto these things, God shall add unto him the plagues that are written in this book: And if any man shall take away from the words of the book of this prophecy, God shall take away his part out of the book of life, and out of the holy city, and from the things which are written in this book.*

(Revelation 22:18,19).

There is also fulfilled prophecy. This requires understanding that the various books of the Bible were written when they claimed to be written. There are thousands of very specific details recorded about hundreds of prophecies. The only way to disbelieve these prophecies is to either believe that the prophecies were written after the fact or to believe that the events prophesied about did not occur the way history records them. As one example, the book of Daniel (He lived and wrote c. 610- c. 520 BC) is so detailed and specific about the prophecies of Persia and Greece that it is mathematically impossible for them to be random chance guesses. Just a few small parts of these prophecies worked out according to the common laws of probabilities and statistics have odds far greater than many trillions to one. These prophecies of Daniel are either divinely inspired or they are after the fact forgeries. There is no other choice. And prophecies with this much detail are scientific evidence. Each statement can be broken down and analyzed according to the laws of probability and statistics.

Though Solomon warns us in Ecclesiastes 12:12 that "of making many books there is no end; and much study is a weariness of the flesh," he still commands us in Proverbs 4:7 to "get wisdom: and with all thy getting get understanding." Nearly one thousand years after Solomon, the Apostle Paul restates this command with "Study to show thyself approved unto God, a workman that needeth not to be ashamed, rightly dividing the word of truth" in 2 Timothy 2:15. While this includes academic or 'book knowledge', it is so much more. "Study to be quiet, and to do your own business, and to work with your own hands, as we commanded you," wrote Paul in 1 Thessalonians 4:11. We must persist in this massive task. As Isaiah said, "line upon line, precept upon precept, here a little, there a little" (Isaiah 28:10 & 13). We will not learn everything in one day, one month or even one year.

Random efforts frustrate us. Without the proper attitude we will waste our time, perhaps even harm others. Jeremiah commanded us to "learn not the way of the heathen" (Jeremiah 10:2). The apostle John commands us to "love not the world, neither the things in the world" (1 John 2:15). According to the Bible, everyone who was ever born can be divided into one of three categories: believer, unbeliever or compromiser.

We must understand, however, what these words mean. Secular Humanists use the words quite differently from the way they are used in the Bible. "How strangely will the Tools of a Tyrant pervert the plain Meaning of Words!"[10] said Samuel Adams. They use the word believe as a synonym for opinion. But the Bible uses the word believe in the sense of a juror accepting someone's testimony as true. Believe means to evaluate the facts and come to a reasoned conclusion. Belief is intellectual understanding combined with the will. The first time the word believe is used in the Bible, Abraham believed God. The next time the word believe is used Israel believed the report that Joseph was still alive. The next time the word believe is used, God is speaking to Moses out of the burning bush. Moses tells God that the elders of Israel will not believe his testimony. God gives Moses three signs and promises that they will believe his testimony. The opposite of belief, unbelief, can be a lack of understanding. But unbelief is usually an act of the will. When the Apostle Paul told the Philippian jailer to believe in the Lord Jesus Christ to be saved (Acts 16:31), the jailer understood what Paul had said and decided to act on what he understood. Paul also wrote to the Thessalonian believers that those who have "pleasure in unrighteousness" will be sent "a strong delusion, that they should believe a lie" (2 Thessalonians 2:11). They chose to have pleasure in unrighteousness. That choice blinded their understanding. Believing the Bible means to intellectually understand that what it says is true, to accept that what it says as true, to acknowledge that it applies to me and to obey it (faith).

Truth is an absolute. For the Bible to be true, it cannot change over time. As the Psalmist said, "For ever, O LORD, thy word is settled in heaven" (Psalm 119:89). As Malachi said,"For I am the Lord, I change not" (Malachi 3:6). Truth does not depend on circumstances. Truth is not different for different people ("That is just *your* truth"). We learn truth through our reason and our five senses, not through our feelings. The word feelings is often used to mean our emotions combined with our desires. The only way we can learn truth though our feelings is when we use the word feelings to mean our combined experiences and senses without having enough information to arrive at a firm conclusion. Though truth does not change, our understanding of truth will change as we mature.

When we believe that the Bible is true, we will take action. Faith is taking action based on our beliefs.

"The just shall live by his faith" (Habbakuk 2:4). Correct faith will display evidences and substance. As the writer to the Hebrews said, "faith is the substance of things hoped for, the evidence of things not seen" (Hebrews 11:1). A synonym for faith is obedience. Partial or incomplete obedience may seem like belief in the beginning, but when testing comes, the evidence of this unbelief is ultimately disobedience. Jesus used stony or rocky places in the parable of the sower and the seeds to illustrate this. These people seem to be believers because they receive the word of God with joy. But when trials come, they "fall away." They believed with their understanding. They chose to believe with their will. But their faith (obedience) only lasted until problems came. Then the rocky soil no longer had faith. That is, the "converts" no longer obeyed. (Matthew 13:18ff)

Another common word used throughout the Bible which secular humanists misuse is trust. Trust is very similar to belief and faith. Trust, however, is confidence placed in someone or something. Trust may or may not be rational. The important part of trust is the object we trust in. "Trust in the LORD with all thine heart; and lean not unto thine own understanding" (Proverbs 3:5). Compared to faith and belief, trust is passive; who or what we trust in does the acting. Trusting in the Lord is similar to resting in or waiting on the Lord. "Neither let Hezekiah make you trust in the LORD, saying, The LORD will surely deliver us: this city shall not be delivered into the hand of the king of Assyria" (Isaiah 36:15, KJV). Hezekiah's great grandfather, Uzziah, fortified Jerusalem, built engines of war on her towers, had a strong military and made Judah a strong kingdom. Assyria, however, overpowered all of these. Hezekiah's trust in the Lord at this time was completely passive. There was nothing he could do.

The biggest, most egregious lie is that these words are special religious words. The truth is that we use these words the same way the Bible uses these words thousands of times each day. When we examine a chair, we will believe either that it can or that it cannot support us. If we sit in the chair, we are exercising faith in the chair. As long as we remain seated in that chair, we are trusting that it will continue to support us. If the chair collapses under us, it might be a funny comedy routine, but it also shows misplaced belief, faith and trust.

Secular humanists also take ordinary words and give them special religious meanings, though they do not use the word "religious." In the Bible and in traditional court proceedings, the word "acceptance" is a legal term similar to adoption. Unbelievers, however, use acceptance in a specialized religious sense to mean that sin is unimportant and should be ignored. This was the message of the pop psychology book, *I'm OK, You're OK*. Believers use acceptance to mean that sin is atoned for through the blood of Jesus Christ. Believers should forgive one another as God, for Christ's sake, forgave us (Eph 4:32). The phrase "accepted in the beloved" in Eph 1:6 is a description of our adoption by God through the atonement of the blood of Jesus Christ. It does not mean, as unbelieving Secular Humanists indoctrinate, that God overlooks and ignores our sin or that there is no God and therefore, no sin.

The most commonly misused word, however, is science. Unbelieving Secular Humanists have elevated their concept of science to a standard of worship. Used properly, the word science means the use of human reason to organize observations. Science is not technology. Technology is the application of this organized knowledge.

> *The LORD hath called by name Bezaleel the son of Uri, the son of Hur, of the tribe of Judah; And he hath filled him with the spirit of God, in wisdom, in understanding, and in knowledge, and in all manner of workmanship; And to devise curious works, to work in gold, and in silver, and in brass, And in the cutting of stones, to set them, and in carving of wood, to make any manner of cunning work. And he hath put in his heart that he may teach, both he, and Aholiab, the son of Ahisamach, of the tribe of Dan. Them hath he filled with wisdom of heart, to work all manner of work, of the engraver, and of the cunning workman, and of the embroiderer, in blue, and in purple, in scarlet, and in fine linen, and of the weaver* ... (Exodus 35:30-35).

Today we think of technology as working in well-equipped research laboratories. Louis Pasteur said, "There does not exist a category of sciences to which we can give the name of 'applied sciences.' There are science and the applications of science."[11] Moses wrote of technology as skills which these men had enabling them to build the tabernacle. While these technological skills are built upon science, technology is not, in and of itself, science. The King James Version of the Bible translates "understanding science" as a skill of Daniel and his three friends in Daniel 1:4. This understanding science was what we would think of as understanding knowledge.

Not all knowledge is true science. Paul wrote to Timothy in I Timothy 6:20 that he should be on guard against and stay away from those who used "oppositions of science falsely so called." This is one of the more interesting references in the Bible to false science. He was telling Timothy to stay away from people who opposed the Word of God with false knowledge. Our lives are brief and we should not waste time studying things which are not true. Since true religion is the foundation or basis of true science, true religion is where our studies should begin if we are to understand true science. The fear of the Lord is the beginning of wisdom (Psalm 111:10). The fear of the Lord is the beginning of knowledge: but fools despise wisdom and instruction. (Proverbs 1:7) The fear of the Lord is the beginning of wisdom: and knowledge of the holy is understanding (Proverbs 9:10).

The following is a basic overview of Christian doctrines. It is not what anyone claims that the Bible says, but what it actually says. These are the foundations of true science. Without these foundations, a person is not studying, teaching or practicing science. He is a believer in a false religion, studying, teaching and practicing mythology.

Cosmology

The material universe had a beginning. God spoke it into existence. God used His Son and His Spirit to make the universe.

In the beginning was the Word, and the Word was with God, and the Word was God. The same was in the beginning with God. All things were made by him; and without him was not any thing made that was made (John 1:1-3).

God, who at sundry times and in divers manners spake in time past unto the fathers by the prophets, Hath in these last days spoken unto us by his Son, whom he hath appointed heir of all things, by whom also he made the worlds; Who being the brightness of his glory, and the express image of his person, and upholding all things by the word of his power (Hebrews 1:1-3).

Through faith we understand that the worlds were framed by the word of God, so that things which are seen were not made of things which do appear (Hebrews 11:3).

In the beginning God created the heaven and the earth (Genesis 1:1).

God is eternal.

Before the mountains were brought forth, or ever thou hadst formed the earth and the world, even from everlasting to everlasting, thou art God (Psalm 90:2).

Lord, thou hast been our dwellingplace in all generations. Before the mountains were brought forth, or ever thou hadst formed the earth and the world, even from everlasting to everlasting, thou art God (Psalm 93:2).

God is holy.

For I am the LORD that bringeth you up out of the land of Egypt, to be your God: ye shall therefore be holy, for I am holy (Leviticus 11:45).

But as he which hath called you is holy, so be ye holy in all manner of conversation; Because it is written, Be ye holy; for I am holy (I Peter 1:15, 16).

God created the material universe very good. That is, without sin (holy) and good for man.

And God saw every thing that he had made, and, behold, it was very good. And the evening and the morning were the sixth day (Genesis 1:31).

God created the material universe out of nothing.

Thus saith the LORD, thy redeemer, and he that formed thee from the womb, I am the LORD that maketh all things; that stretcheth forth the heavens alone; that spreadeth abroad the earth by myself (Isaiah 44:24).

The portion of Jacob is not like them: for he is the former of all things; and Israel is the rod of his inheritance: The LORD of hosts [armies] is his name (Jeremiah 10:16)

The portion of Jacob is not like them: for he is the former of all things; and Israel is the rod of his inheritance: The LORD of hosts is his name (Jeremiah 51:19)

All things were made by him; and without him was not any thing made that was made (John 1:3).

God that made the world and all things therein, seeing that he is Lord of heaven and earth, dwelleth not in temples made with hands; Neither is worshipped with men's hands, as though he needed any thing, seeing he giveth to all life, and breath, and all things (Acts 17:24,25).

Through faith we understand that the worlds were framed by the word of God, so that things which are seen were not made of things which do appear (Hebrews 11:3).

Anthropology (The Study of Man)

And God said, Let us make man in our image, after our likeness: and let them have dominion over the fish of the sea, and over the fowl of the air, and over the cattle, and over all the earth, and over every creeping thing that creepeth upon the earth. So God created man in his own image, in the image of God created he him; male and female created he them (Genesis 1:26-27).

Man was created by a direct act of God.

And the LORD God formed man of the dust of the ground, and breathed into his nostrils the breath of life; and man became a living soul (Genesis 2:7).

Then the LORD God said, "It is not good for the man to be alone; I will make him a helper suitable for him" (Genesis 2:18, NASB, NIV).

And the LORD God caused a deep sleep to fall upon Adam, and he slept: and he took one of his ribs, and closed up the flesh instead thereof; And the rib, which the LORD God had taken from man, made he a woman, and brought her unto the man (Genesis 2:21, 22).

Man was created without sin.

And God saw every thing that he had made, and, behold, it was very good. And the evening and the morning were the sixth day (Genesis 1:31).

And they were both naked, the man and his wife, and were not ashamed (Genesis2:25).

By one man sin entered into the world, and death by sin; and so death passed upon all men, for that all have sinned (Romans 5:12).

Sin entered the universe through Adam's choice.

And the LORD God took the man, and put him into the garden of Eden to dress it and to keep it. And the LORD God commanded the man, saying, Of every tree of the garden thou mayest freely eat: But of the tree of the knowledge of good and evil, thou shalt not eat of it: for in the day that thou eatest thereof thou shalt surely die (Genesis 2:15-17).

And they heard the voice of the LORD God walking in the garden in the cool of the day: and Adam and his wife hid themselves from the presence of the LORD God amongst the trees of the garden. And the LORD God called unto Adam, and said unto him, Where art thou? And he said, I heard thy voice in the garden, and I was afraid, because I was naked; and I hid myself. And he said, Who told thee that thou wast naked? Hast thou eaten of the tree, whereof I commanded thee that thou shouldest not eat? And the man said, The woman whom thou gavest to be with me, she gave me of the tree, and I did eat. And the LORD God said unto the woman, What is this that thou hast done? And the woman said, The serpent beguiled me, and I did eat. And the LORD God said unto the serpent, Because thou hast done this, thou art cursed above all cattle, and above every beast of the field; upon thy belly shalt thou go, and dust shalt thou eat all the days of thy life: And I will put enmity between thee and the woman, and between thy seed and her seed; it shall bruise thy head, and thou shalt bruise his heel. Unto the woman he said, I will greatly multiply thy sorrow and thy conception; in sorrow thou shalt bring forth children; and thy desire shall be to thy husband, and he shall rule over thee. And unto Adam he said, Because thou hast hearkened unto the voice of thy wife, and hast eaten of the tree, of which I commanded thee, saying, Thou shalt not eat of it: cursed is the ground for thy sake; in sorrow shalt thou eat of it all the days of thy life; Thorns also and thistles shall it bring forth to thee; and thou shalt eat the herb of the field; In the sweat of thy face shalt thou eat bread, till thou return unto the ground; for out of it wast thou taken: for dust thou art, and unto dust shalt thou return. And Adam called his wife's name Eve; because she was the mother of all living. Unto Adam also and to his wife did the LORD God make coats of skins, and clothed them (Genesis 3:8-21).

The result of Adam's sin is death.

By one man sin entered into the world, and death by sin; and so death passed upon all men, for that all have sinned (Romans 5:12).

The curse of sin is on the entire material universe.

For we know that the whole creation groaneth and travaileth in pain together until now (Romans 8:22).

Because we are born sinners, we are powerless to heal the curse of sin.

For all have sinned, and come short of the glory of God (Romans 3:23).

We have before proved both Jews and Gentiles, that they are all under sin; As it is written, There is none righteous, no, not one: There is none that understandeth, there is none that seeketh after God. They are all gone out of the way, they are together become unprofitable; there is none that doeth good, no, not one. Their throat is an open sepulchre; with their tongues they have used deceit; the poison of asps is under their lips: Whose mouth is full of cursing and bitterness: Their feet are swift to shed blood: Destruction and misery are in their ways: And the way of peace have they not known: There is no fear of God before their eye (Romans 3:9-18).

Christology (The Study of Jesus Christ)

Jesus was born of a virgin according to the prophets.

Now all this was done, that it might be fulfilled which was spoken of the Lord by the prophet, saying, Behold, a virgin shall be with child, and shall bring forth a son, and they shall call his name Emmanuel, which being interpreted is, God with us (Matthew 1:22,23).

Jesus the Messiah was born without sin.

For he hath made him to be sin for us, who knew no sin; that we might be made the righteousness of God in him (2 Corinthians 5:21).

Jesus the Messiah led a sinless life.

Seeing then that we have a great high priest, that is passed into the heavens, Jesus the Son of God, let us hold fast our profession. For we have not an high priest which cannot be touched with the feeling of our infirmities; but was in all points tempted like as we are, yet without sin (Hebrews 4:14,15)

God provided atonement for the sin of all mankind through the death, burial and resurrection of Jesus Christ.

For since by man came death, by man came also the resurrection from the dead. For as in Adam all die, even so in Christ shall all be made alive (I Corinthians 15:21,22)

Time

The existence of the material universe is both finite and brief. The genealogies in Genesis are consistent and clearly teach that the earth was created shortly before 4,000 BC. The year 2010 in a modern western calendar corresponds to the Jewish calendar year 5770, counted from Creation. Jesus Christ will return and the material world will be destroyed as a judgment for sin.

Heaven and earth shall pass away, but my words shall not pass away. Matthew (24:35).

But the day of the Lord will come as a thief in the night; in the which the heavens shall pass away with a great noise, and the elements shall melt with fervent heat, the earth also and the works that are therein shall be burned up (2 Peter 3:10).

And I saw a great white throne, and him that sat on it, from whose face the earth and the heaven fled away; and there was found no place for them (Revelation 20:11).

And I saw a new heaven and a new earth: for the first heaven and the first earth were passed away; and there was no more sea (Revelation 21:1).

Any who do not believe these simple, obvious doctrines willingly bring judgment on themselves. Like those on the *Titanic* who refused to board the lifeboats, they will not understand reality until it is too late. As Peter said, they "willingly are ignorant."

Knowing this first, that there shall come in the last days scoffers, walking after their own lusts, And saying, Where is the promise of his coming? For since the fathers fell asleep, all things continue as they were from the beginning of the creation. For this they willingly are ignorant of, that by the word of God the heavens were of old, and the earth standing out of the water and in the water: Whereby the world that then was, being overflowed with water, perished: But the heavens and the earth, which are now, by the same word are kept in store, reserved unto fire against the day of judgment and perdition of ungodly men (2 Peter 3:3-7).

Unlike the record of Jesus the Messiah, the idea that the universe had an Intelligent Designer has three forms of testimony outside the Scriptures. The first two are scientific. First is the fact that no basic law of physics contradicts anything in the Bible. The second testimony is the millions of observed scientific phenomena. The last form of testimony consists of other ancient records.

True science is not technology. Technology is the application of scientific principles to fabricate material objects. No known aspect or instance of technology contradicts any doctrine in the Bible.

True science is the belief (accepting as true after examining the facts, a legal term) that the material world was created in a point of time, that the material world will end, that each and every person has a purpose and will be judged for the decisions he makes in this life. True science acknowledges the limitations of material measurements and analysis. True science acknowledges that science is finite and that the most important human questions cannot be answered by science. True science acknowledges that the ancient world of 5,000 years ago was totally and completely different than the world we know today. The present is not the key to the past. True science acknowledges that the ancient world underwent at least one worldwide global catastrophe within the last 5000 years.

True science acknowledges that the surface of the earth that we see today was produced catastrophically. True science acknowledges that each historic document from the beginning of time was written for a purpose at the time it was written. Though technology evolves, such as the evolution of the airplane, there has never been any evolutionary development of civilization. True science acknowledges that the only legitimate use of the word evolution is technological, driven by human intelligence, such the evolution of the automobile, the printing industry or organized sports. True science acknowledges the scientific fact that there has never been any scientific evidence for any form of biological evolution.

1 Sam Harris, *The End of Faith,* W.W. Norton & Company New York, NY, 2004.

2 Andy Rooney, *Sincerely, Andy Rooney*, Essay Productions, Public Affairs, by the Perseus Group, New York, NY, 1999.

3 User Willette, *Askville.Amazon.com,* posted 2009 or early 2010.

4 "*Titanic*" article, *New World Encyclopedia.org*.

5 Flavius Josephus, *The Antiquities of the Jews,* Translator: William Whiston, 1737.

6 "The Date of Christ's Birth," *The Moorings.org*.

7 Eusebius, *Church History or Ecclesiastical History* (Hist. Ecc viii 2.) Written by in the 4th century. Translated by Arthur

Cushman McGiffert, from *Nicene and Post-Nicene Fathers, Second Series,* Vol. 1. Edited by Philip Schaff and Henry Wace. Christian Literature Publishing Co., Buffalo, NY, 1890.

8 University of Oxford, Bodleian Philosophy Faculty Library, Manuscripts and Rare Books "Medieval Manuscript Sources and Incunabula." *ox.ac.uk*.

9 *Biblefacts.org*

10 letter to John Pitts, January 21, 1776.

11 Correspondence I, p. 382-383, "To the Rector of the Academia de Douai," 15 Nov. 1855. Cuny, H., *Louis Pasteur, The Man and his Theories,* Translated P. Evans, London, The Souvenir Press, 1965.

3. Does Science Conflict with Religion?

"I am quite conscious that my speculations run quite beyond the bounds of true science."[1]

Charles Darwin

"In the space of one hundred and seventy-six years the Mississippi has shortened itself two hundred and forty-two miles. Therefore ... in the Old Silurian Period the Mississippi River was upward of one million three hundred thousand miles long ... seven hundred and forty-two years from now the Mississippi will be only a mile and three-quarters long. ... There is something fascinating about science. One gets such wholesale returns of conjecture out of such a trifling investment of fact."[2]

Mark Twain

For hundreds of years before the rise of Secular Humanism, science to the western mind had two parts. The first part is the observation and measurement of the material universe. Even most non-westerners universally accepted this part. Eventually this organization of observations and measurements became known as the scientific method. Though the scientific method took centuries to develop, and the road to development was quite difficult, the scientific method became the most valuable tool in modern scientist's toolbox.

While no two people, organizations, dictionaries or encyclopedias today have exactly the same definition of the scientific method they all include the following basics.

First, state a problem. Second, develop a method of observing, measuring and/or testing. Third, based on initial observations, measurements and tests, develop a hypothesis. Fourth, expand testing, observations and measurements to test your hypothesis. Fifth, research the work of others to see how their measurements, tests and observations affect your hypothesis. Always make your hypothesis fit the facts. Modify your hypothesis to fit the research. If the hypothesis is far enough out of line, scrap your hypothesis and create a new hypothesis more in line with the facts. Sixth, retest the new hypothesis. Repeat these steps as necessary until the hypothesis is in line with the observed data. Then publish your findings and allow others to check your work.

The second part of the definition of science is the organization of theories into a coherent whole. The explanations of these observations and measurements, however, have never had a consensus. One famous example from hundreds of years ago had one side convinced that Ptolemy was correct and that the sun revolved around the earth. Others took the same observations and measurements of the heavens above, sided with Copernicus and concluded that the earth revolved around the sun. Both positions had access to the same basic information, but each side arrived at very different conclusions. Even though both sides came to different conclusions, honest scientists on both sides agreed on one very important aspect of science. The measurements, experiments and scientific observations would determine the outcome.

The difference between these two views of the universe was their interpretation of facts. Their interpretations were based on their preconceptions as well as their misconceptions. The only hope of agreement between these points of view is a willingness to submit to an honest evaluation of the facts. Of course, since scientific professions had some dishonest men, there were conclusions published which best fit the desires of the authors, rather than supporting the facts. The following sad examples taint the hard work of the armies of honest, dedicated and hard-working men throughout the centuries.

Ernst Haeckel, 19th century biologist and artist, is famous for illustrating hundreds of life forms and was considered an expert at recreating sea life and other animals on paper. “As a would-be scientist ... his hand as an artist altered what he saw with what should have been the eye of a more accurate beholder,”[3] said Jane M. Oppenheimer, PhD in embryology and Professor of Biology and History of Science at Bryn Mawr, of Haeckel. He was a theoretical evolutionist and many of the animal forms he illustrated in support of his theories have never been discovered to exist in reality, as Oppenheimer and others pointed out. “He was more than once, often justifiably, accused of scientific falsification, by Wilhelm His (Swiss anatomist and embryologist) and by many others.”[3] Though the Christian (and Jewish and Muslim) ethical standards found this dishonesty a form of extreme evil, nevertheless some of these dishonest works were published and gathered followings.

Dr. J. S. Weiner's book, *The Piltdown Forgery*, chronicles the deliberate fakery involved in the discovery of a famous "missing link." Weiner was a Reader in Physical Anthropology at Oxford University and a committed evolutionist, exposing the fact that this link was no link at all. "On December 18, 1912, Arthur Smith Woodward and Charles Dawson announced to a great and expectant scientific audience the epoch-making discovery of a remote ancestral form of a man--The Dawn Man or the 'Piltdown man.'... a veritable confirmation of evolutionary theory."[4] Sir Arthur Keith, anthropologist and paleontologist, expressing his support for the discovery at the time, spoke in reverent terms about his religion's dogma. "That we should discover such a race, as Piltdown, sooner or later, has been an article of faith in the anthropologist's creed." Weiner reports that "the objective evidence for the deception was overwhelming...The Piltdown 'men' were forgeries, the tools were falsifications, and the animal remains were planted."[4]

Another sad episode in the history of science is the one time belief that slavery was beneficial. Scientists once claimed they had evidence to support the belief that slavery was good for blacks. "In the past, everything from craniometry to phrenology to I. Q. tests were used by scientists to prove the inferiority of non-white races,"[5] writes Suzanne Currie. Currie studied Physics and Biology and graduated from York University.

"Scientists obtained their ideas on race from society; they then proved these ideas using scientific 'facts;' the scientists then presented their proofs to society, thus reinforcing the racist beliefs of that society."[5]

Currie's article makes the point that scientists are normal people who are influenced by their society and its thinking and beliefs as much as anyone is. "It is an unending cycle that still continues in the field of science in the present time. Science became a purveyor of race and racism because scientists are not totally objective and are influenced by the beliefs of society."[5] These influences can color their judgment and even affect the way they conduct their experiments. "They are human beings, subject to emotions, thoughts, and opinions. The racist views of society are reflected in a scientist's work."[5]

"If a scientist is set in his beliefs that one race is inferior to another, and he sets out to prove that idea, he will probably obtain data from his experiment proving that there is a superior race."[5] Not if he is honest! "A scientist cannot simply 'turn off' their personal beliefs and emotions, becoming like a machine, and being totally objective."[5] If he is not "totally objective" then he has no right to call himself a scientist. "Even if the actual data does not prove the hypothesis, the manipulation of the data will."[5] Currie explains the circular reasoning of these supposed applications of the scientific method, reinforcing the almost universal reliance today by evolution supporters on presupposition. "The scientist can eliminate 'unsuitable data,' adjust for certain conditions that he has perceived, and plot data on a graph in a way that will make small differences seem larger."[5]

Though this idea of proving slavery to be scientifically beneficial seems insane and abhorrent to Westerners in the 21st century, this belief was so thoroughly entrenched in the nineteenth century that nations went to war over it. This is, however, only one of many proofs that incorrect scientific conclusions will produce unreasonable men who will not change their beliefs, no matter how much new scientific data they are shown. Anglican minister and essayist Sydney Smith said,"Never try to reason the prejudice out of a man. It was not reasoned into him, and cannot be reasoned out."[6]

Most of the errors and erroneous conclusions in early scientific works, however, were not due to dishonesty, but due to either poor scientific equipment or poor scientific methods. Greater discipline throughout the scientific community helped to greatly reduce obviously faulty research. More scientists helped to verify results. The scientific method gradually became standardized. The erroneous conclusions were more difficult to deal with. It quickly became obvious that adults who were ingrained with faulty scientific conclusions might never accept the fallacy of the conclusions they held so dear. So education of children is of critical to change the thinking of an entire population. Although changing incorrect scientific beliefs, such as the belief that men would never fly, is very difficult, the American, British and German education systems of the late eighteenth and early nineteenth centuries were very successful at doing just that. They were helped by the extremely rapid advances in technology which allowed men to see the error of their ways.

In 1804 Richard Trevithick invented a type of locomotive that ran on roads rather than rails. He was among the first of what later became the steam engine. "I have been branded with folly and madness for attempting what the world calls impossibilities."[7] He says as he describes the resistance of other inventors and scientists of his day. "And even from the great engineer, the late Mr. James Watt, who said to an eminent scientific character still living, that I deserved hanging for bringing into use the high-pressure engine."[7] Trevithick could only take comfort in the knowledge that his work would be vindicated one day because he knew it was based in truth. "[I have] ... been the instrument of bringing forward and maturing new principles and new arrangements of boundless value to my country."[7]

But the greatest help in overcoming incorrect scientific beliefs was support by churches. "All religions, arts and sciences are branches of the same tree,"[8] said Albert Einstein. "All these aspirations are directed toward ennobling man's life, lifting it from the sphere of mere physical existence and leading the individual towards freedom. ... It is no mere chance that our older universities developed from clerical schools."[8] From this statement it is still clear that he did not see true religion and true science as adversaries but as united in the goal of bettering man's existence. "Both churches and universities — insofar as they live up to their true function — serve the ennoblement of the individual."[8]

Instead of building on this sure foundation, most modern scientists rely on "ancient wisdom" such as that of Egypt and form their theories around these early works, denying that the source of all truth is in God and the Scriptures. Just as the Bible records pagans robbing the Temple of its treasures to dedicate to their false gods, these "scientists" steal truths that can be observed in nature and use them to adorn their preconceived temples of falsehood. This can, however, become a two way street.

"I am stealing the golden vessels of the Egyptians to build a tabernacle to my God from them, far far away from the boundaries of Egypt,"[9] Kepler boasted in contrast to these pseudoscientific pagans. He had learned these ancient sources also but his studies led him to defy the idolatry of preconceived knowledge. "If you forgive me, I shall rejoice; if you are enraged with me, I shall bear it."[9] He hoped he could convert the "heathen" who misused science. But he realized they might attack him because his discoveries were for the glory of God. "See, I cast the die, and I write the book."[9] Kepler humbly left the matter of truth finding its audience in God's hands.

"Whether it is to be read by the people of the present or of the future makes no difference: let it await its reader for a hundred years, if God himself has stood ready for six thousand years for one to study him."[9] At the same time, he believed that it was man's duty to present truth and not continue to build on thousands of years of flawed human reasoning that excluded God.

German physicist Walter Heitler strongly agreed with Kepler, insisting that "A contradiction (between science and religion) is out of the question. What follows from science are, again and again, clear indications of God's activity which can be so strongly perceived that Kepler dared to say (for us it seems daring, not for him) that he could 'almost touch God with his hand in the Universe'."[10] Scientific discovery once thrived among believers in Christ. But there were also churchmen who allowed their worship of man's reason to stifle their ability to recognize truth and overcome preconceptions just as secularists refuse to accept truth today. "In the long run my observations have convinced me that some men,"[11] Galileo scoffed, "reasoning preposterously, first establish some conclusion in their minds which, either because of its being their own or because of their having received it from some person who has their entire confidence, impresses them so deeply that one finds it impossible ever to get it out of their heads."[11] Galileo knew that such ideas didn't originate from truth but from relying on reason or society's influence.

"Such arguments in support of their fixed idea as they hit upon themselves or hear set forth by others, no matter how simple and stupid these may be, gain their instant acceptance and applause."[11] It is clearly blinding pride that makes this mindset, Galileo observes. "On the other hand whatever is brought forward against it, however ingenious and conclusive, they receive with disdain or with hot rage — if indeed it does not make them ill."[11] Modern secularists express their opposition to Scriptural truth in exactly these terms. "Beside themselves with passion, some of them would not be backward even about scheming to suppress and silence their adversaries."[11] Nothing has changed since Galileo's time.

Galileo's conflict with the Church over the Heliocentric Theory was by no means as clear a case of truth versus censorship as later interpreters present it. But a part of the controversy lay with those who had accepted the Aristotelian teaching that an immovable Earth was the center of things. They re-interpreted Scriptures that did not even deal with cosmology to support this ancient humanist doctrine. As Isaac Newton later said, "Plato is my friend — Aristotle is my friend — but my greatest friend is truth."[12] Galileo's frustration was due in part to the wrong position of the Church and its taking upon itself the sole right to interpret Scriptures, not some conflict in Scriptures themselves with his discoveries.

"Natural Philosophy" was a term frequently used as synonymous with natural science. Part of the earliest duty of the church was to try to make certain truth prevailed over error. The Established Churches oversaw the early university studies to try to prevent the breeding of heresies as they branched out into different areas of study. Some scholars wanted to keep the emerging natural sciences studies separate from theological studies but the church wanted to oversee what was being disseminated. This system failed because the Church became hopelessly corrupt and heretical itself. "Later Philosophers banish the Consideration of such a Cause out of natural Philosophy,"[13] Isaac Newton said, condemning the attempt to separate physical causes from spiritual ones. He accused these scholars of "feigning Hypotheses for explaining all things mechanically, and referring other Causes to Metaphysicks..."[13]

That is, they let their preconceived notion of the need to separate the studies dictate false hypotheses that could explain every natural phenomenon. Anything spiritual they claimed lay outside the realm of science or natural philosophy. "Whereas the main Business of natural Philosophy is to argue from Phenomena,"[13] that is, the purpose of the study of natural science is to draw conclusions from observation, "without feigning Hypotheses, and to deduce Causes from Effects..."[13] In other words, Newton demanded study of what could be known to exist, not the concoction of theories about what might be imagined, "till we come to the very first Cause, which certainly is not mechanical."[13] Thus Newton reconnected science and religion and claimed that anyone who was separating them was "feigning" or inventing what was not true.

Sir Francis Bacon, saw clearly the direction scientific study was headed in its quest to separate itself from religion. "It is true, that a little philosophy inclineth man's mind to atheism, but depth in philosophy bringeth men's minds about to religion. ...For while the mind of man looketh upon second causes scattered, it may sometimes rest in them, and go no further."[14] Bacon meant that superficial study lead to an inability to understand the source of all things. Bacon, however, stated that his goals were the discovery of truth, service to his country, and service to the church. "But when it beholdeth the chain of them confederate, and linked together, it must needs fly to Providence and Deity."[14] He believed atheism resulted from not studying in sufficient depth and that looking deeper to see how things were connected led to acknowledgement of God.

Secular Humanists, however, rely on people like John William Draper and Andrew Dickson White. “White was the most famous and successful exponent of the conflict hypothesis. ...”[15] The conflict hypothesis teaches that science and religion are concerned with different things. It relies on 19th century research “proving” that religion and science should either exist independently or be antagonist toward one another. “It is worth briefly examining whether White was being entirely honest in his work as no one doubts that Draper was engaged in nothing more than polemic...”[15] James Hannam, PhD in the History and Philosophy of Science from the University of Cambridge, gives background on the fabrication of the “conflict” between science and religion.

Concerning Draper and White, Hannam says, “Neither of them were professional historians and both did seem to sincerely believe in the warfare theory they were expounding.”[15] There is nothing like sincerely false belief to color a researcher’s judgment. “Unfortunately, this meant that they set out to prove what they already believed rather than take their conclusions from the facts.”[15] Hannam takes note of White’s reliance on feelings rather than facts. “White is quite explicit about this when he writes how he felt before he began his research, ‘I saw... the conflict between two epochs in the evolution of human thought - the theological and the scientific’.”[15] This highly biased, untrained author is one of the commonly used sources of the secularist argument that theists are violent opposers of discovery and new learning. "... He [White] thought that in the Middle Ages especially, the Church was burning freethinkers left, right and centre."[15]

Hannam does not charge White with fabricating data. "One would like to take the charitable view that White really believed his theory and was not making up evidence to support a position he knew to be false."[15] He tries to give him the benefit of the doubt. "Instead, he skews the evidence by accepting that which agrees with his hypothesis while being sceptical of what does not."[15] Unfortunately, the best that can be said about White's data was that it was wholly based on presupposition and designed to prove entrenched ideas. "This means that he has included falsehoods that he would have noticed if he had taken a properly objective attitude towards all his evidence."[15]

Though this bad, it is even worse that these same falsehoods and erroneous conclusions are still taught as facts in classrooms today. These falsehoods are necessary links to go back to a time to when the Church of Rome in the name of Christianity censored or banned books. They ignore the fact that monasteries preserved learning during the time known as the Dark Ages. Nor is the fact that Roman Emperors censored, banned and burned books by the tens of thousands mentioned. Yet, one of the favorite points secularists bring up is the practice of censoring or banning books in the Middle Ages by Rome. "[In 1277] ...many theses derived from Aristotle and Averröes were declared heretical both at Paris and at Oxford following a papal sponsored investigation by Bishop Stephen Tempier."[15] Most Christians, loyal to Rome or not, opposed these declarations at the time. Tainting any Christian of the past four hundred years as a "book burner" is another dishonest secularist claim.

Hannam also explains that these condemnations sometimes included works on conflicting sides of the issues. This resulted in a "double truth" heresy, which allowed a religious truth separate from secular truth. This "double truth" heresy, a familiar concept for today's secularists, has been condemned universally for over five hundred years. "The Averroists had allegedly tried to insist on the doctrine of the double truth whereby philosophy and theology were kept in separate boxes but this was roundly condemned."[15] The separation of "philosophy" (frequently including the early scientific studies) from theology, far from being promoted, was considered a heresy because it involved creating separate definitions for truth.

Specifically, Thomas Aquinas was assigned to "sanitizing" the works of Aristotle to make them accessible to Christians. Yet some of Aquinas's works devoted to that task were also condemned at this same time. "Scholars" like White take no notice when supposed heretics are later installed as bishops and their "heresies" pronounced free of error. The whole purpose is to discredit the system of cooperative learning conducted by the churchmen, scholars and universities of the Middle Ages.

"In general, there was religious support for natural science by the late Middle Ages and a recognition that it was an important element of learning. The extent to which medieval science led directly to the new philosophy of the scientific revolution remains a subject for debate, but it certainly had a significant influence."[15]

A number of more modern scientists strive to maintain the integrity of their profession in the face of much dishonesty on the part of committed Secular Humanists. University of California Professor of Psychology Stanley Sue believed that it was essential to avoid the common secularist redefining of the word "theory" into "fact," as Richard Dawkins frequently does when speaking of Evolution. Sue instead demanded that the bias so evident in secularist dogma be avoided.

> *"Scientific skepticism is considered good. ... Under this principle, one must question, doubt, or suspend judgment until sufficient information is available. Skeptics demand that evidence and proof be offered before conclusions can be drawn. [...] One must thoughtfully gather evidence and be persuaded by the evidence rather than by prejudice, bias, or uncritical thinking."*[16]

Richard Feynman, 20th century physicist, apparently had no use for skewed data and the common practice of simply burying contradictions to theories being researched. He cautioned scientists to be thoroughly honest in their work by including supporting and contrary evidence and anything discovered along the way that might advance knowledge.

> *"If you're doing an experiment, you should report everything that you think might make it invalid — not only what you think*

> *is right about it; ... You must do the best you can — if you know anything at all wrong, or possibly wrong — to explain it. ... Those things it fits are not just the things that gave you the idea for the theory; ...The idea is to try to give all of the information to help others to judge the value of your contribution; not just the information that leads to judgment in one particular direction or another."*[17]

When scientists ignore this commitment to honesty, they fall into the same trap that Isaac Asimov did. Claiming to speak as a scientist, he rightly invoked the word "inspired" to express his baseless but religiously held beliefs. "We can make inspired guesses, but we don't know for certain what physical and chemical properties of the planet's crust, its ocean, and its atmosphere made it so conducive to such a sudden appearance of life..."[18] Although he appears to be humbly admitting science's limitations, Asimov is in fact dishonestly claiming that when a scientist guesses, it is like ordinary people stating facts. Notice that instead of allowing for the possibility of a creative act by God, he assures us that all that happened was a "sudden appearance" of life made possible by natural conditions.

In the novel *Pride and Prejudice*, an unscrupulous man plays on social prejudices to advance his own position just as many secularists advance their "scientific" theories. He pretends humility while providing supposed evidence for theories people already hold. Jane Austin said, "Nothing is more deceitful ... than the appearance of humility. It is often only carelessness of opinion, and sometimes an indirect boast."[19] Those who misuse science frequently advance the "scientist's" own reputation without presenting sound science or true knowledge. "Carelessness of opinion" is almost a watchword for those who feel free to advance any belief and call it science, and frequently they receive applause when they should be greeted with healthy skepticism.

Today the observations and measurements of the physical world must support the established religion of Secular Humanism. "Carelessness of opinion" expressed by their celebrity pseudo-scientists along with their "inspired guesses" must be given as much weight as facts. Its adherents of course, deny this. They loudly denounce the corruption of the Church-State union and insist they are pure of such entanglements. John W. Draper, 19th century American physician and photochemist, claimed that "Science has never sought to ally herself with civil power. She has never subjected anyone to mental torment, physical torment, least of all death, for the purpose of promoting her ideas."[20] While asserting that theists are invariably corrupt and violent, he deflects attention from the hand-in-glove relationship of secularists with the courts resulting in the bombarding of schools and government institutions with lawsuits demanding removal of any hint of religious mention in the name of "separation of church and state."

Thomas Henry Huxley,"Darwin's Bulldog," gave secularists a plan to follow for dealing with conflict between religious and scientific subjects. "The antagonism between science and religion, about which we hear so much, appears to me purely factitious, fabricated on the one hand by short-sighted religious people," he maintained. Huxley believed that it was narrow-minded to argue about the issues from either side. His premise sounds suspiciously like the Averoists above. He sneered at both those "who confound ... theology with religion; and on the other by equally short-sighted scientific people who forget that science takes for its province only that which is susceptible of clear intellectual comprehension."[21] This means that there is obviously nothing intellectual about religious belief and "true" scientists can smugly stand aloof from such petty bickering.

---

1 Charles Darwin, "Letter to Asa Gray," (Harvard Professor of Biology), 18 June, 1857.

2 Mark Twain, *Life on the Mississippi,* first edition published by Osgood and Company, 1883.

3 Jane M. Oppenheimer, *Essays in the History of Embryology and Biology* pp. 153-154, 1967.

4 J. S. Weiner, *The Piltdown Forgery,* Oxford University Press, London, 1955.

5 Suzanne Currie "Science and Racism." *Science Lives! Online,* August 26, 2007.

6 Sydney Smith, 1771 –1845, "Elementary Sketches of Moral Philosophy," delivered at the Royal Institution 1804-1806, published 1850.

7 Richard Trevithick, "Letter to Davies Gilbert," 1804.

8 Albert Einstein, "Moral Decay" (1937); Later published in *Out of My Later Years* (1950), Philosophical Library, New York, and Estate of Albert Einstein, 1956, 1984.

9 Johannes Kepler, *Harmonices Mundi* ("Harmonies of the Earth"), Book Five (1618). Translated E J Aiton, A M Duncan, J V Field, *The Harmony of the World, Memoirs of the American Philosophical Society*, Philadelphia, 1997, 209.

10 Walter Heinrich Heitler, 1904 –1981, *Man and Science* (German title: *Der Mensch und Die Naturwissenschaftliche Erkentnis*), translated by Robert Schlapp, Oliver and Boyd, Edinburgh, 1963.

11 Galileo Galilei, *Dialogue Concerning the Two Chief World Systems,* Italian version 1632, translated by Stillman Drake, University of California Press, 1953 (revised 1967). Also Modern Library paperback.

12 Isaac Newton, Personal Notes in Latin titled: *Quaestiones Quaedam Philosophicae ("Certain Philosophical Questions")* (c. 1664). A. Rupert Hall and Marie Boas Hall (eds. and trans.), *Unpublished Scientific Papers of Isaac Newton* (1962, reissued 1978).

13 Newton, *Medieval World Laid the Foundations of Modern Science,* with related articles entitled, "Medieval Science and Philosophy" and "Science and Church in the Middle Ages," Icon Books, London, 2009.

16 Stanley Sue, "Science, Ethnicity, and Bias: Where Have We Gone Wrong?" *American Psychologist,* 54 (12), December 1999.

17 Richard Feynman, “Cargo Cult Science,” adapted from the Caltech commencement address given in 1974.

18 Isaac Asimov, *Omni,* November 1983, p. 58.

19 Jane Austin, *Pride and Prejudice,* 1813.

20 John W. Draper, author of *History of the Conflict between Religion and Science,* 1874.

21 Thomas H. Huxley, “The interpreters of Genesis and the interpreters of Nature,” 1885.

4. What Does The Scientific Evidence Prove?

"I am quite conscious that my speculations run quite beyond the bounds of true science."

Charles Darwin[1]

"Facts are stubborn things; and whatever may be our wishes, our inclinations, or the dictates of our passion, they cannot alter the state of facts and evidence."

John Adams[2]

"In the space of one hundred and seventy-six years the Mississippi has shortened itself two hundred and forty-two miles. Therefore ... in the Old Silurian Period the Mississippi River was upward of one million three hundred thousand miles long ... seven hundred and forty-two years from now the Mississippi will be only a mile and three-quarters long. ... There is something fascinating about science. One gets such wholesale returns of conjecture out of such a trifling investment of fact."

Mark Twain[3]

Samuel Johnson said, "Integrity without knowledge is weak and useless, and knowledge without integrity is dangerous and dreadful."[4] Christians often believe that science is an enemy. They think this way because most who use the word "science" have completely abandoned Johnson's demand that integrity go hand-in-hand with knowledge. Uniformitarians replace truth with selective evidence which supports preconceived conclusions. Christians should neither develop an antagonism toward true science, nor should they ignore the very real contributions of true science. True scientists should not ignore the very real foundation of science in Christianity. Dennis Prager, author, columnist, radio show host and historian, laments the unthinking reliance on pseudo-science in today's society.

> *"In much of the West, the well-educated have been taught to believe they can know nothing and they can draw no independent conclusions about truth, unless they cite a study and 'experts' have affirmed it. 'Studies show' is to the modern secular college graduate what 'Scripture says' is to the religious fundamentalist."*[5]

In everyday life, probably the greatest area of conflict between Christianity and those who misuse the word science is moral relativism. This came about because of the dishonest use of the word relativity. The theories of relativity (special and general) neither support nor have any reference to moral relativism. The similarity in the sound of the words is simply a propaganda technique. Neither do the moral absolutes of the Word of God belittle true science. However, since the established religion of Secular Humanism teaches just the opposite, the following list of scientific facts can help us understand that true science can only be explained by the world as described in the Bible. None of these scientific facts prove the Bible. Each does prove the religious belief in Deep Time to be a scientific impossibility.

Sun Energy Source

Until shortly after WWII, the majority of scientists believed the sun shrank by an average of .01 percent per year. They concluded, based on 400 years' worth of scientific observations, that this was proof that the sun was powered by gravitational collapse. In the late 1800s the Kelvin–Helmholtz Contraction Theory was developed to explain both the observed contraction and how the sun was powered.[6]

In the 1930s religious Secular Humanists, understanding the consequences of this theory, proposed nuclear fusion as the energy source for the sun. In 1928 George Gamow published a paper proposing a theory. The Gamow theory included what came to be known as the Gamow factor.[7] This was the first serious step toward the idea that stars are powered by nuclear fusion. General acceptance was slow. Gamow's ideas were further explored in the 1930s. This exploration grew into Hans Bethe's theory of Stellar Nucleosynthesis.[8] Bethe won a Nobel Prize for this work in 1967. The detonation of the atomic bombs actually did more to convince people than the theoretical papers. According to the Kelvin-Helmoltz Contraction Theory, with gravitational collapse powering the sun, the sun would have been so large and hot around 50,000 years ago that it would have boiled all the water out of all the oceans on the earth. A mere million years ago the orbit of the earth would have been inside the sun.[9]

These are not just creationist rantings. While we might disagree about the exact numbers, there is no questioning the basic principle. Here are the words of very committed uniformitarians.

"But Kelvin–Helmholtz contraction cannot be the major source of the Sun's energy today. If it were, the Sun would have had to be much larger in the relatively recent past. Helmholtz's own calculations showed that the Sun could have started its initial collapse from the solar nebula no more than about 2.5 million years ago. But the geological and fossil record shows that the earth is far older than that, and so the Sun must be as well. ***Hence, this model of a Sun that shines because it shrinks cannot be correct.***"[10]

Why is the Kelvin-Helmholtz contraction incorrect? Because they *believe* "that the earth is far older than that."

Secularists' religious belief that life on earth has existed for billions of years required the sun to be billions of years old. So they eagerly accepted that the sun was powered by nuclear fusion. It is also true that the sun's surface is not static, as is the case with water, rock, or ice on a planet's surface, so a consistent measurement of the sun's diameter is not possible. Two different measurements on the same day might yield different diameters because of massive fluctuations on the sun.

The sun's life cycle in this theory would be 9.2 billion years. At 4.6 billion years the sun would be roughly half way through its life cycle. The problem with this theory is that the sun would have been too cool; forty percent of its present brightness, 4.6 billion years ago, and the earth would have been a frozen wasteland. This is also known as the early faint sun paradox.

Project *SOHO*, launched in 1995, has provided us with more information about the sun. Unwilling to change their belief about the age of the sun and life on earth, some Secular Humanists now believe that the sun is powered by a perfect balance of different types of nuclear fusion combined with gravitational collapse, which has produced a uniform temperature for life on earth for the last 4.6 billion years. The best way to describe this kind of balance is "miraculous."[11]

NASA's Solar Dynamics Observatory Satellite, or SDO, was launched in February 2010 and chief scientist Dean Pesnell said it has already reshaped our theories of how the star works.[12]

ScienceBlogs.com contributor Ethan Siegal, PhD in theoretical astrophysics at the University of Florida, writes a blog called "Starts with a Bang." In a post titled, "How the Sun works, from the inside out." originally written August 12, 2011 he has included crossed out and updated information, indicating how rapidly information changes. Theories about the sun must be changed, updated, and corrected to match the most up-to-date information.[13]

Many who insist on billions of years agree that 4.6 billion years is just a rough estimate. They explain that the 9.2 billion year figure is correct, but that they simply are not certain how far our sun is into its 9.2 billion year lifecycle. With such rapid changes in information, how is it possible to state dogmatically that the sun has a lifespan of 9.2 billion years?

Radiohalos and Radiometric Dating

The 4.6 billion year figure comes from radiometric dating. The most common radiohalos are found in zircons. Zircons are found all over the earth and range in size from microscopic to the size of rocks. Zircons are crystals of zirconium silicate. Zircons can be of gemstone quality and large zircons are often used as substitutes for diamonds (cubic zirconia is a close man-made synthetic). Most zircons, however, are around the size of very small grains of sand. Though the wide range of impurities creates an enormous variety of zircons, the most important geological use for zircons is radiometric dating. Zircons contain trace amounts of uranium and thorium. Tiny zircons are sliced open to examine fission tracks produced by decaying uranium. These fission tracks and the U→Pb (Uranium to lead) methods of radiometric dating have dated zircons as old as 4.404 billion years.[14] This means, according to uniformitarian assumptions, that the original crystal formation of these tested zircon samples occurred 4.404 billion years ago. As with all radiometric samples, there is no scientific way of knowing the original condition of the sample tested.

The same zircons also contain helium, which provides another dating method. Helium retention rates in these zircons have a date range beginning as young as 4,000 years.[14] While there are a tremendous number of articles attacking this date as too young, most of these articles read like tabloids reporting on British royalty, not honest science. A true scientist would look for an explanation that accounted for both the uranium radiohalos and helium retention. It is far easier to come up with an explanation for the radiohalos' presence if the rocks have a young age. It is nearly impossible to explain the retained helium if the rocks are ancient. One possible explanation for young radiohalos is that they were created at the same time the earth was created.

Another possible answer is that a massive thermonuclear event such as the sun exploding and losing an outer layer sent thermonuclear radiation through the entire planet. Since such an event would have destroyed all non-aquatic life on earth, this would have happened around 2350 BC, at the time of the flood.

Whatever the trigger, the most likely scenario is accelerated nuclear decay during the flood. The accelerated nuclear decay during the flood is the only explanation we are aware of which properly balances all the existing evidence.

Dating Rock Formations

Travel anywhere and you will see signs proclaiming various rocks or layers of rocks to be millions or even billions of years old. "The Little Willow Formation consists primarily of contorted quartz schist and gneiss; at 1.7 billion years old, it is the oldest rock in the Salt Lake City area."[15] This religious belief in the myth of great ages is proclaimed in textbooks, school classrooms, museums, and, most importantly, by employers, who will either refuse to hire or actually fire anyone who refuses to openly support this religious myth. Though many supposed scientific facts are used to support these dates, radiometric dating is the foundation for all dating methods.[16]

Since fossils have no direct methods of dating, such as radiometric dating, how is the age of a particular fossil determined? It is compared to other fossils of a "known" age or it is dated according to strata where it was found. How is the age of the other fossils or the strata determined? The usual answer is more fossils of a "known" age or strata of a "known" age. This circular reasoning is standard textbook content. Radiometric dating is the only dating method that does not rely on a comparison with something else already dated. Radiometric dating is not valid on fossils. Radiometric dating must be used on the rocks in the surrounding strata.

Every form of radiometric dating always depends on an unstable radioactive isotope deteriorating (decaying) to a stable isotope at a known, stable rate. K→Ar (Potassium to Argon) and U→Pb (^{235}U to Lead) are just two examples of isotopes used as radiometric clocks. All radioactive isotopes used as radiometric clocks work the same way. A sample is taken and analyzed. The result is always a ratio of radioactive isotope to the stable final element. The ratio can be found on a chart because the rate of decay is stable, that is, unaffected by any known outside influence.

Everyone who uses radioactive isotopes as clocks must, however, rely on three invalid religious assumptions. The first is the unscientific religious leap of faith that the sample being tested had no daughter material when it was formed. For radiometric clocks to work, it is an absolutely essential scientific necessity to know the original condition of the sample. If the original condition of the sample is less than one hundred percent radioactive, then the date is less than the published date. These widely publicized dates, however, assume that the tested sample began its existence (at the time it was formed) one hundred per cent radioactive with no daughter element present. In reality, the only scientific information we learn from these radiometric tests is the current ratio of radioactive material to nonradioactive material. The published dates are always the upper possible date of a range of possible dates.

The second assumption is that nothing contaminated the sample during its existence. It is impossible to know the conditions of the sample during its existence. Simply to assume that contamination never occurred is not science. It is especially bad science when the very people claiming the accuracy of radiometric dating also claim that the earth underwent a series of what they call extinction events which catastrophically changed the surface of the earth.

The third assumption is that nothing changed the decay rate. While the second assumption is something changed the individual sample or the immediate environment of the sample, this assumption is worldwide or throughout the universe. In other words, is the present really the key to the past? How can we know that?

If there was a sample with a known date, other samples could be calibrated against it. Not only do such calibration samples not exist, without a time machine to go back in time and obtain a calibration sample, it is not possible to have a calibration sample. With very few exceptions, the published dates of tens of thousands to billions of years are based purely on mythological religious assumptions.

There are a few exceptions, however. Thermoluminescence is used to date pottery. Thermoluminescence is based on the time the pottery was fired, a known event. ^{14}C is another one of these exceptions. The radioactive isotope ^{14}C is integrated into the tissue of living organisms throughout the lifetime of that organism. At the time of death that amount of ^{14}C becomes a fixed amount and begins to radioactively decay into ^{14}N. That amount of ^{14}C at the time of death is equal to the amount of ^{14}C in the atmosphere at the time of death. The ratio of ^{14}C in the sample tested at the time of death compared to the assumed amount of ^{14}C in the atmosphere at the time of death should be an almost perfect radiometric clock. However, certain conditions can alter that ratio. If the ratio is in some way tampered with, that tampering is normally leaching. That particular sample will then test to be older than it really is. ^{14}C dates older than 1000 B.C. also assume atmospheric amounts of ^{14}C and absorption rates of ^{14}C similar to current absorption rates. That is almost certainly an invalid assumption, making many, if not all ^{14}C dates for items dated older than 1000 B.C. older than they actually are.

Finally, radiometric dating never gives an absolute date. It only tells us the upper limit of a range of possible dates. One of the most important but rarely published conclusions by secularists about ^{14}C dating is that even they admit that after approximately 60,000 years, the remaining ^{14}C is such a small amount that it is difficult to test for it.

They have also concluded that ^{14}C disappears entirely after a little more than 100,000 years. Though fossils which are pronounced older than 100,000 years are rarely tested for ^{14}C, those few which have been tested usually have some trace of ^{14}C.

"Carbon-14 (^{14}C) dating of multiple samples of bone from 8 dinosaurs from Texas, Alaska, Colorado, and Montana revealed that they are only 22,000 to 39,000 years old."

"After the AOGS-AGU conference in Singapore, the abstract was removed from the conference website by two chairmen because they could not accept the findings. Unwilling to challenge the data openly, they erased the report from public view without a word to the authors or even to the AOGS officers, until after an investigation. It won't be restored.

The researchers presented their findings at the 2012 Western Pacific Geophysics Meeting in Singapore, August 13-17, a conference of the American Geophysical Union (AGU) and the Asia Oceania Geosciences Society (AOGS)."[17]

Lunar Recession

"Scotty, I need power now!" pleads a desperate Captain Kirk, on *Star Trek* the original series, only to hear his Chief Engineer reply, "He's turned the engines off. Completely cold. Thirty minutes to startup. I canna change the laws of physics, Captain!" The only hope of pulling out of a decaying orbit and saving the *Enterprise* is a restart of her engines.

Why? The Earth is surrounded by nonpowered satellites. Why does the *Enterprise* need power?

The *Enterprise* needs to remain in a slow orbit as near the planet as possible while maintaining a fixed point over it. The only way to accomplish that is to keep the engines powered up to constantly change the orbit and counter the pull of gravity. The closer a satellite is to the object it is orbiting, the faster it has to travel to maintain a stable orbit without engine power. Mercury is the fastest-traveling planet orbiting the sun. Venus, Earth and Mars are all progressively slower at their greater distances.

For a given velocity there is only one stable orbit around (distance from) a planet (or sun). The closer the satellite (or starship) is to the object it is orbiting, the faster it needs to travel (without power or thrust). Otherwise it needs power (thrust) to maintain the orbit.

All orbiting objects have one of three types of orbits.

1) The orbit is a decaying orbit (being pulled into the planet),

2) A stable orbit (no change in orbit over time) or

3) A receding orbit (moving away from the planet).

The closer the satellite (or starship) is to the object it is orbiting, the faster it must travel or the more power (thrust) it must use to counteract the attraction of gravity.

The laser reflectors left by the Apollo missions on the lunar surface have been hit with lasers from earth many times. It proves the moon is moving away from the earth at a rate of about 1.5 inches per year.[18] Assuming the Moon has not changed mass, velocity or direction of orbit, sometime in the past it would have possessed a stable orbit closer to the

Earth. Even closer to the earth, inside that stable orbit, the moon's orbit with no change in mass, velocity, or orbital direction, would have decayed.

The important point is the stable or equilibrium orbit.

At the current rate of recession, the moon is receding from the earth at a rate of approximately one mile every 42,240 years. If the Moon were to remain unchanged in mass, velocity and rate of recession, just over 42 million years ago, the moon would be more than one thousand miles closer to the earth. With no change in its mass and velocity, in far less than 42 million years the moon's orbit would be a decaying orbit, not a receding orbit. The moon would be drawing closer to the Earth and not moving away from the earth as it is now. Just like the *Enterprise,* the moon would have a decaying orbit.

Since the lunar orbit is receding, not decaying, it means that the existing lunar orbit is, geologically speaking, very young. If you reject the conclusion that this indicates a recent creation, you are left with only two choices:

1) The Moon is a very recent addition to the Earth or

2) Some sort of catastrophe has relatively recently altered the Moon's orbit. This catastrophe would have required far more energy than what could be produced by the simultaneous detonation of all the nuclear weapons on Earth.

That catastrophic event would have occurred quite recently, even according to an Uniformitarian point of view. Any catastrophe powerful enough to alter the moon's orbit would also have catastrophically affected the Earth as well.

Geomagnetic Field Decay

The earth's magnetic field seems to have a half-life of around 1,400 years. Assuming that to be correct, just 11,200 years ago the earth's magnetic field would have been 256 times stronger than it is now. 14,000 years ago the earth's magnetic field would have been 1,024 times stronger than it is now. That would make animal life impossible.[19]

Dr. Russell Humphreys has continued to work in this field. He believes that a strict linear half-life is unlikely and concludes his article with the following: "These ideas weigh heavily against the idea that there is currently a 'dynamo' process at work in the core that would ultimately restore the lost energy back to the field. Without such a restoration mechanism, the field can only have a limited lifetime, in the thousands of years. So the clarity of this new fit, especially the exponential part, is further evidence that the earth's magnetic field is young."[20]

The *dynamo process* Dr. Humphreys references is the theory that instead of continually decaying, earth's magnetic field is continually rebuilding or restoring itself. Its existence would make the entire life cycle of the earth's magnetic field much more stable and therefore much, much older.

Many articles purport to "debunk" these observations. They usually begin with personal attacks on the original authors. Even if the "debunkers" are correct that the half-life of the earth's magnetic field is longer than 1,400 years, for life to exist hundreds of millions years ago, the half-life must be hundreds of times larger. Instead of making personal attacks, detractors should present data supporting a much longer half-life. Instead, they use data based on the presupposition that the earth is millions of years old to insist that the earth's magnetic field must have a large half-life.

An excellent response to these uniformitarian proposals is Jonathan Sarfati's article, *The Earth's Magnetic Field: Evidence That the Earth Is Young.* After including many detailed examples, it concludes, "The clear decay pattern shows the earth could not be older than about 10,000 years."[21]

A key element of the decay of the earth's magnetic field is the evidence of magnetic field reversals found in cooled lava flows. "Lava flows in Nevada's Sheep Creek Range may have preserved evidence that the planet's magnetic field rapidly changed direction." Other evidence in the paleomagnetic lava flow at Steens Mountain in Oregon shows that the earth's magnetic field reversed its polarity in just over two weeks. This is in line with a magnetic field half-life of 1,400 years, but entirely unexplainable with a magnetic field hundreds of millions or billions of years old. Also, the sun reverses the polarity of its magnetic field every eleven years. [22]

Mount Saint Helens Canyon Formation

Though the eruption of this volcano was observed by millions via television, internet and print media, few are aware of the significance. It carved another "Little Grand Canyon" similar to the Grand Canyon in Arizona in days or perhaps only hours. Millions witnessed this event. We saw it happening live or on video in real time. To those who understand, the evidence is so overwhelming that even some evolutionists are admitting it, though quite grudgingly. They still believe, without any evidence, that the catastrophe creating most other prominent geologic sites happened millions of years ago. But the evidence has forced them to admit that the Grand Canyon's formation was rapid, created with massive amounts of water.[23]

Varves

Paint brush shows scale.

Photo Information/Credit[24]

Mount Saint Helens produced more scientifically observable evidence, in the form of something remarkably like varves. According to most dictionaries, a varve is a layer of mud at the bottom of a pond or lake. Counting varves used to be considered a very important method of dating. Each layer has two parts, one lighter and the other darker. One part is the time of year when water moves quickly and brings many deposits. The other part is the time of year when water is stationary, perhaps even stagnant, such as the winter when the lake is covered in ice. All honest scientists admit that there might be more than one varve per year. A core sample, which counts varves to determine the age of a lake, might give a date older than the actual date. The assumption that conditions in the past were similar to the present is an assumption that the difference in dates is slight. Though the varve count might be slightly older than the actual date, it is still assumed to be reliable.

The eruption of Mount St. Helens on May 18, 1980 scientifically proved that assumption to be in error.

Ash and mud up to twenty-five feet thick with thousands of layers filled the region. Though these layers are never called varves, they display the same characteristics as varves, a thin light and dark layer sandwiched together. These thousands of layers, put down in a matter of hours, perhaps minutes, completely discredit the concept of using layers for dating. People who use varves for dating never acknowledge or even mention this.[25]

Lake Titicaca

Catastrophes change the shape of the land where they occur forever. Few places have been changed more dramatically than Lake Titicaca. It rises 12,500 feet (3,800 meters) above sea level in South America between Bolivia and Peru. Fresh water rivers and streams feed the lake now, but the salt content is about five-and-a-half parts per thousand, classifying the lake as brackish. No current conditions explain the salt content of the lake. The most likely explanation is that Lake Titicaca was once at or below sea level and that a catastrophic uplift moved a lake full of seawater almost two and a half miles above sea level. Marine fossils are preserved around the lake. The current salt level of Lake Titicaca is about fifteen percent of the Pacific Ocean off Peru, where the Pacific Ocean varies from 34 to 37 parts per thousand.

"In the heart of the Andes ... is Lake Titicaca." [Its saltiness and the present fauna (It contains sea horses)] strongly suggested that the present fauna of Lake Titicaca has survived from a time when the lake communicated directly with [was connected to] the ocean."[26]

Theories other than uplift have been postulated. These theories, however, require completely speculative processes that have never been observed. That is not a scientific process. Some evolutionists believe the uplift theory because the scientific evidence is so overwhelming, but say that the catastrophe occurred millions of years ago. A past waterline is slanted in relation to the current waterline.

At one time Lake Titicaca had more water than it does now, and this waterline is much higher at one end of the lake, proving that a catastrophe tipped the lake in the past.

"The strandline [near Lake Titicaca] was carefully surveyed for a length of about 375 miles [603 km]. And then it was established that it is not 'straight.' ... Its level showed a slant of a most peculiar character in relation to the present ocean-level, or, which amounts to the same, relative to the present level of Lake Titicaca."[27]

Past irrigation and buildings in and around the lake prove an ancient civilization built the port city of Tiahuanaco. It is approximately 800 feet higher than the current lake surface. Tiahuanaco was a harbor, but is now twelve miles south of the lake, and the ship berths are of a size most likely built for ocean-going vessels. The lake "is large, 3,261 square miles. It contains sea horses, suggesting that this region or its water were once below sea level. On the mountain sides ... are terraces of ancient corn fields going up to 17,000 feet. Yet corn will not germinate [sprout] above 11,500 feet!"[26] There is also a building beneath the lake, about 660 feet long, with a road running to it and roads and steps leading down into deeper water. It is twice the size of a modern soccer field and modern archeologists believe it to be a temple.[28]

Himalayan Yellow Band Ammonites

Geological features suddenly and catastrophically created by water can be found all over the world. Near the top of the Himalayan Mountains, reaching up to 29,029 feet (8,848 m) above sea level, is a layer of rock known as the yellow band. This yellow band is filled with marine fossils called ammonites. "Marine fossils are also found high in the Himalayas, the world's tallest mountain range, reaching up to 29,029 feet (8,848 m) above sea level. For example, fossil ammonites (coiled marine cephalopods) are found in limestone beds in the Himalayas of Nepal. All geologists agree that ocean waters must have buried these marine fossils in these limestone beds."[29] Ammonites are common fossils found all over the earth and are similar to a modern marine creature known as the Nautilus. These fossilized marine creatures lived in oceans, not a freshwater lake. All of the Himalayan mountain chain had to have been under seawater. This many fossilized ammonites were not transported somehow after the mountains were formed. Sometime in the past either the entire Himalayan Mountain chain was five miles lower than it is now or the entire planet had more than twice the water volume than it has now.

Also, these ammonite fossils are not crushed. They were moved by a catastrophic (sudden) event while the entire layer was plastic (mud). The ammonites had to be protected to keep them from being crushed. The only scientific answer is that the surrounding mud protected them. A slow uplift over a long period of time would have dried out the surrounding mud. Instead of protecting the ammonites, the hardened rock would have ground the ammonites to powder. Since water weighs approximately eight pounds per gallon, the energy necessary to create the Himalayan Mountains was thousands of times greater than all the nuclear weapons in all the nuclear arsenals in every country on earth detonating at the same time.

On a much smaller scale, highway engineers have carved into mountains all over the earth. Thousands of mountains at every possible elevation all over the earth show bent, folded and twisted layers intact without cracks. These layers had to be plastic [mud] when they were put in place. This is the only way these mountains could have formed.

Catastrophic Fossil Formation

Fossils like the ammonites of the yellow band were formed in a catastrophic event. Even evolutionists have to admit that. Insects, microbes and weathering combine to destroy everything except bones in less than one hundred years, often in less than a week. Tissue deteriorates too rapidly for fossilization without a catastrophe. The question is: When did this catastrophe occur? Once again, the only available scientific evidence indicates a more recent event. Since fossils were bones, they once contained ^{14}C. If a standard test for ^{14}C discovered any ^{14}C, the tested sample would scientifically prove a date less than 60,000 years old.

Since the religion of Secular Humanism demands that these fossils are millions of years old, and they control both the fossils to be tested and the testing procedures, ^{14}C testing is never done on fossils. There is only the dogmatic assertion that fossils are now stone and there is no carbon to test.

There have been carbon tests on mollusks that indicate a much younger date on certain samples. Without ^{14}C testing, however, there is no scientific dating of fossils, only religious pronouncements.[30]

Warm Antarctica

There is abundant evidence that Antarctica was once warm. Warm climate animal and plant fossils abound. Antarctica also has a coal bed. Formation of a coal bed requires a massive number of warm climate plants. The ice core samples have discovered that many of these fossils are below sea level.[31] The question is, why? Possible explanations include that Antarctica was once in a warmer latitude and moved, or that the magnetic poles have shifted (as mentioned earlier), or that the oceans have changed dramatically. The facts are clear. A once warm-climate land mass now lies beneath ice three miles thick in spots. The common thread through all of these explanations is the scientific fact of a catastrophic change. Sadly, instead of completely scientific explanations, supposed scientific and academic articles lead off with dogmatic mythological assertions: "around 40 million years ago," "millions of years ago," "40,000 year cycles."

A catastrophe could turn Antarctica from a warm climate into a frozen wasteland very quickly. The articles describing Antarctica's past might have much correct scientific data, but these dogmatic mythological foundational assertions make the entire article read like something from a tabloid.

Ice Core Samples and *Glacier Girl*

Researchers annually take core samples of glaciers and massive ice caps such as those covering Antarctica. They count each layer as a single year and publish a date. As with radiometric dating, it is not possible to know the original condition of the ice cap or glacier. However, unlike radiometric dating, starting with a zero condition (no ice, bare ground) is reasonable. Assuming that the earth is older than the glacier or ice cap is also reasonable. So the speed with which a glacier or ice cap forms is critical to an accurate date.

A squadron of P-38s made an emergency landing on a glacier in Greenland on July 15, 1942. The pilots were rescued and the planes abandoned. On July 15, 1992, the final piece of one P-38 was dug out of the glacier. For 50 years the squadron of planes had flowed with the glacier while being covered with more ice. One plane was pulled out of the glacier more than two miles from the landing point, buried in 268 feet of ice. The recovered plane was restored and renamed *Glacier Girl*.[32]

If the same techniques which boldly pronounce with absolute certainty that core samples are tens of thousands of years old were applied to the ice which covered *Glacier Girl*, then that ice would have to be thousands of years old. Yet the scientific fact is that *Glacier Girl* was covered by 268 feet of ice. It is also a scientific fact that this ice was fifty years old. This particular glacier grew at a rate of almost 5.5 feet per year. At this rate, the thickest ice sheet in the world, Terre Adelie in Antarctica, could be formed in just over 2,500 years. Secular Humanists would insist that different conditions in the past required a much longer formation time. Any change in the conditions could just as easily have resulted in a shorter time for formation.

Besides counting the layers in core samples, methods of dating a glacier include testing the oxygen content of the water, studying the presence of CO_2, finding evidence of radioactive decay, and examining pieces of volcanic material. Fred Hall wrote a brief article, "Ice Cores Not All That Simple,"[33] showing how complicated glaciers really are. None of these methods are entirely reliable. While there have been hundreds, perhaps thousands, of computer and theoretical models developed to explain glacier formation, the only scientific observations are in line with the glacier which covered *Glacier Girl.*

Water Scale

Formations like the Grand Canyon were carved by water action. No one disputes this, but water action is a topic on which scientists can still find points of disagreement.

Anyone who has watched a movie from the 1950s or earlier has seen the common, at that time, technique of filming a scale model ship in a tank of water. Even with rather large ships, such as the ones used in Ben Hur, the water never looks quite real. The explanation is that water does not scale, that is, large amounts of water look and behave differently than smaller amounts of water. For realistic water, producers and directors had to either build full sized models or wait decades for 3-D animation.

Since water does not scale visually, scale models using water were believed to be inaccurate also. For example, water can cut channels through layers of mud. The resulting canyons look something like a scale model of the Grand Canyon. These models have been dismissed as unrealistic simply because water does not scale. This idea that water does not scale, however, is not always true. Bridges built in powerful rivers, especially rivers prone to flooding, are often washed away. Engineers building a new bridge over the Mississippi River at Alton, IL, between 1990 and 1994, built a scale model to help prevent that from happening. The scale model they built used real water to examine the direction and amount of force the Mississippi River would exert on the new bridge.[34]

Though the water did not "look" to scale, the water flow and pressure information was accurate enough to allow the engineers to see what forces they had to contend with. Because of the model, modifications were made in the bridge's design. These modifications are likely the reason the bridge is still standing today.

Water Salinity

The study of water and its properties gives scientists many ways to learn about the earth. It is also a very much-abused study when it comes to dating the earth. One of the most unusual methods of dating is measuring the salt content of the sea. This method assumes a constant and steady addition of salt to the sea by freshwater rivers and a constant and steady evaporation rate.[35]

While this method is often cited in textbooks as one of many dating methods "proving" an ancient earth, it is fraught with so many difficulties that it has few educated defenders. It is usually referred to as an "additional" or "support" method of dating.

The first major scientific problem with using the salt content of the sea as a dating method is the wide disparity in salt content among the world's bodies of water. The Mediterranean Sea has a much higher salt content than the Atlantic, which has a higher salt content than the Antarctic. Is the Mediterranean older than the rest of the oceans of the world? Are the rivers that flow into the Mediterranean saltier than the rest of the rivers of the world? The scientific answer to both of these questions is either "no" or "we do not know".

People have attempted to date the seas using salt content since at least Sir Isaac Newton's time. Simply measuring the contact of salt requires a date no more than 100 million years old. Drs. "Austin and Humphreys calculated that the ocean must be less than 62 million years old. It's important to stress that this is not the actual age, but a maximum age. That is, this evidence is consistent with any age up to 62 million years, including the biblical age of about 6,000 years."[36]

Providence Canyon

Water's power to shape the landscape provides us with another amazing creation in southwest Georgia. Providence Canyon, sometimes called "Georgia's Little Grand Canyon" is 1100 acres of canyons carved out by erosion since the early 1800s.[37] Though much smaller than the Grand Canyon, it is still one of the largest canyons in America. As the model of the Alton, Illinois bridge helped engineers understand the Mississippi River; Providence Canyon helps us understand the Grand

Canyon in Arizona. While neither the Alton Bridge nor the Providence Canyon are perfect models, on a small scale they both accurately model the effects of far greater amounts of water.

Though Providence Canyon is smaller than the Grand Canyon, it has many similar features. The formation of Providence is a scientific fact. Since Providence Canyon is less than two hundred years old, there is no scientific reason to believe the Grand Canyon, or any other canyon, must be millions of years old. All that is needed to make a canyon larger than Providence Canyon is more water, not more time.

Arches National Park

Just a few hundred miles north of the Grand Canyon is Arches National Park in Southeast Utah. When the National Park Service took over management of the park in 1971 there were over 2000 natural sandstone arches. Since then many arches have collapsed, the most famous being the collapse of Wall Arch on August 4-5, 2008. Though the exact number of collapsed arches is not documented, if only one arch were to collapse every three years, in less than 675 years there would not be any arches left. There is no evidence that there were ever hundreds of thousands or millions of arches. The evidence indicates that there were never many more arches than we see now. And there is evidence that many arches have fallen since 1971.

The scientific conclusion, based on the scientifically observed collapse of arches, is that either conditions in the recent past were drastically different to preserve the arches or that the arches are much younger than Secular Humanism proclaims them to be. If conditions were similar to the conditions we know now, the arches could be not older than about 4,000 years old. They could easily be younger.[38]

Stalactites and Stalagmites

Photo Credits[40]

Stalactites and stalagmites are another geologic feature formed by water. The Lincoln Memorial was built between 1914 and 1922 over land formerly at the bottom of the Potomac River. Stalactites and stalagmites have formed beneath the stairs in the period since the construction was completed (less than 100 years at the time of this writing) as described below.

"... Supports were driven down almost 100 feet to bedrock. This cavernous foundation provides an ideal location for cave-like formations to actively grow. Water on the surface slowly makes its way down through cement stairs into the empty space below. Cement is 'glued' together by calcium carbonate. As the water passes through the cement, it dissolves some calcium carbonate and carries it downward. When the water reaches the open space, it leaves behind the calcium carbonate, creating stalactites and stalagmites. This is the same way cave stalactites and stalagmites form, only instead of passing through cement stairs, natural cave features usually form when ground water passes through limestone. Since the cave formations at the Lincoln Memorial aren't in an actual cave, maybe we should call them 'under-the-stair-ites'."[40]

Dave E. Matson attempts to claim that all the stalactites and stalagmites in existence could not have formed in less than 5000 years.

"The Bulletin of the National Speleological Society (37: p.21, 1975) gives a stalactite/stalagmite growth rate of ... 0.1 to 10 centimeters per thousand years. An exceptional spurt of growth might exceed ... 10 centimeters or 2.5 inches per thousand years. ...Thus, a 60 foot giant, as might be found in Carlsbad Caverns, would have a minimum estimated age of about 180,000 years."[41]

Timpanogos Cave National Monument, in American Fork, Utah, is described in park literature as being roughly 65 million years old. Although the claim that the stalactites and stalagmites are millions of years old is no longer being made, secularists still demand thousands of years for the formation of these cave features, usually tens or hundreds of thousands, not the less than a hundred necessary in the Lincoln Memorial.

"Water trickling through the limestone overlying the caves dissolved calcite and other minerals from the rock. ...The water deposited its mineral load as tiny crystals on a cave ceiling, wall, or floor. Over thousands of years, as countless crystals were deposited, a variety of cave formations took shape--stalactites, stalagmites, flowstone, helictites, and others." [42]

Julia E. Cole, University of Tucson professor of Geosciences, studied a limestone cave in southern AZ. The article described the cave formations as "natural climate archives" and said, "The stalagmite yielded an almost continuous century-by-century climate record spanning 55,000 to 11,000 years ago. ... Each climate regime lasted from a few hundred years to more than one thousand years."[43] The article's photo shows Sarah Truebe, a geosciences doctoral student, with a cave formation less than 2 feet tall. The cave itself appears to be less than ten feet in height. The use of these cave features as "natural climate archives" is based on utter conjecture and uniformitarian presupposition.

Each of these claims is nothing more than a gratuitous assertion. The evidence shows the formation of these features, especially of stalactites, take only decades.[44]

Diamonds

Diamonds are reported by Secular Humanists to be at least 45 million years old, formed 75 miles or deeper beneath the earth. Yet laboratory-produced diamonds are made commercially. In 1953 the first industrial grade laboratory-produced diamonds were made. These tiny laboratory-produced diamonds were and still are manufactured for industrial applications such as cutting tools and electrical coatings. More advanced processes coat lenses with laboratory-produced diamond. The laboratory produced diamond industry gradually increased the size and quality of laboratory-produced diamonds. In early 2010 a 2-carat laboratory-produced gemstone-quality diamond was produced in about three days.

The FTC (Federal Trade Commission) requires the label of laboratory-produced, laboratory-manufactured, or for the product to be identified by the name of the manufacturer along with the specific type of gem rather than using the term synthetic when referring to the type of gem. A laboratory-produced diamond is identical to a geologically-produced diamond is every way;

chemically, structurally, and appearance, even on the microscopic level.

A synthetic diamond is still a technically correct term when applied to a laboratory-produced diamond, but some vendors use the term "synthetic" diamond to refer to cubic zirconia, cut glass or any other manmade substitute for a diamond. So the FTC wants to distinguish between real diamonds, both geologic and laboratory-produced, and other crystals which are not diamonds. These other crystals which are not diamonds are often marketed as synthetic diamonds, which they are not.

The existence of laboratory-produced diamonds is a scientific fact. They are produced and reproduced under controlled conditions. Depending on color desired, other chemicals can be added to the crystal-growing process and the resulting gemstone can be pink, green, blue, yellow, and colorless (white). The diamond mining industry is concerned about the number of laboratory-produced diamonds being manufactured. Proclaiming the youngest naturally-occurring diamond to be 45 million years old is a dogmatic religious pronouncement contrary to science. The most reasonable scientific theory is that diamonds were formed suddenly under the correct conditions.[45]

The RATE (Radioisotopes and the Age of The Earth) research project at the Institute for Creation Research found Carbon 14 in both natural diamonds and ten US coal beds. The upper range of the possible dates for the diamonds was 55,000 years.[46]

"R.E. Taylor of the Department of Anthropology at the University of California-Riverside and of the

Cotsen Institute of Archaeology at the University of California-Los Angeles teamed with J. Southon at the Keck Accelerator Mass Spectrometry Laboratory of the Department of Earth System Science at the University of California-Irvine to analyze nine natural diamonds from Brazil. All nine diamonds are conventionally regarded as being at least of early Paleozoic age, that is, at least several hundred million years old. So if they really are that old they should not have any intrinsic ^{14}C in them. Eight of the diamonds yielded radiocarbon 'ages' of 64,900 years to 80,000 years."[46]

Coal and Oil

Carbon can be turned into diamonds, coal and oil. Most of the world uses coal and oil to heat homes and run electric generators. Coal and oil are abundant. Coal was made from plant material that was covered with water and neither coal nor oil is being formed geologically today. There is, however, one possible exception.

When Mount St. Helens erupted in 1980, the explosion filled Spirit Lake with logs. The bottom of the lake was filled with plant material covered in mud from the eruption. This was under the lake water. Examination of this material showed the early stages of the formation of coal. This, however, is a very tiny deposit compared to the massive coal deposits worldwide.[47]

Crude petroleum is also produced on a small scale in laboratories from garbage. Like diamonds and petrified wood, the processes that formed coal beds and oil deposits are not going on today. Something is dramatically different about conditions today. Like diamond formation and petrified wood formation, the only possible scientific example of oil and coal formation shows coal and oil being formed rapidly in a catastrophic event.[48]

Petrified Wood

Just off of Interstate 40 in the Navajo Indian reservation in northeast Arizona is the Petrified Forest. While the Petrified Forest in Arizona is the most well-known of the Petrified Forests, petrified wood can be found all over the world, though it is abundant in dry regions of the western United States. Petrified wood is wood that has turned to stone. The authors have seen petrified tree stumps in South Dakota more than twenty feet in diameter. Perhaps there is a rare exception somewhere, but wood today is not becoming petrified except synthetically. Though modern wood can be synthetically turned into petrified wood, the process only works under highly-controlled conditions.

The existence of synthetic diamonds and manmade petrified wood both scientifically prove that these substances require not immense periods of time but proper conditions to form them. They can be made in days, weeks or months. The religious belief that wood petrified over a great period of time and that the process took place thousands or millions of years ago is without scientific evidence. Even people who insist that wood petrified thousands or millions of years ago must admit wood is not naturally petrifying today. Therefore, something has dramatically changed. Though the most reasonable scientific explanation is a catastrophe, religious dogmas refuse to allow for these scientific explanations.[49]

Japanese scientists fastened pieces of wood in a hot acid lake. They observed that "after only 7 years the wood had turned into stone, petrified with silica."[50]

Preservation in Ice, Tar, and Peat

There is an entire class of creatures entombed in tar, peat bogs, muck and sand; organic remains clearly younger than fossils. Some of the more well-known of these creatures are woolly mammoths, mastodons, saber-toothed tigers, woolly rhinos, and giant ground sloths. These creatures are not fossilized, that is, turned to stone, but frozen. Though some were preserved in warmer areas by tar, most specimens are frozen and decay quickly when thawed.

Undigested plant remains in their stomachs prove that the animals were frozen quickly. Though pop culture places these animals in the Ice Age, the types of undigested plants found in their stomachs suggest that at the time the animals were frozen they and the plants lived in a warm climate. This indicates that tropical, subtropical or at least temperate conditions existed even in Siberia, Alaska, and northern Canada. Though these creatures obviously died in a catastrophe, this catastrophe was far more recent than the catastrophe which fossilized billions of other creatures worldwide.[51]

Soft Tissue in Dinosaur Bones

Creatures found in tar and ice are said by evolutionists to have come from a later time period than those fossilized, thus minimizing the troublesome issues of their real age. It is harder to dismiss one discovery, soft tissue from the bone of a dinosaur. There is no question that soft tissue deteriorates very rapidly. There is no question as to what such a discovery must mean. Dinosaurs, at least the ones leaving remains with soft tissue, had to have been living at the same time humans lived. The question is, has such a discovery actually been made? Paleontologist Mary Higby Schweitzer of North Carolina University has published her discovery of the remains of blood cells in dinosaur fossils and soft tissue remains in a T. Rex. John Asara of Harvard Medical School and researchers at Palo Alto have verified it.

Blood for transfusions can only be stored for about six weeks. Cryopreservation is only good for about ten years. These facts should help put this discovery into perspective. Not only did Dr. Schweitzer's team find bone cells, blood cells, and proteins, they also found dinosaur DNA.[52]

We should also compare this to the soft tissue discovered in a female wooly mammoth carcass. Even uniformitarians only date this to be 10,000 years old. We know that this is more than twice the actual age of the wooly mammoth. The conditions for preservation of the tissue were much better. It was preserved in permafrost at -10° C. Physiologist Kevin Campbell of the University of Manitoba wrote in an email to Kate Wong in a May 30, 2013 article that, "ancient DNA is highly fragmented and by no means "ready to go" into the next mammoth embryo." He also said "how were these samples preserved in this state for so long?"[53]

So the soft tissue of Ice Age woolly mammoths preserved in very favorable conditions is puzzling.

The preservation of dinosaur soft tissue for millions of years is, by comparison, impossible.[54]

The evidence is overwhelming. The conclusions, however, are even more astounding. Rather than admit the obvious, that the bones with the remains of soft tissue were young, the published reports maintain that the bones are millions of years old, but some process not now understood preserved the soft tissue. There is no evidence of such an unknown process. This is a classic attempt to make the evidence fit the preconception.[54] It is also a glaring example of professional bigotry. This kind of article can never be published without a claim that the bones are millions of years old.

Mary Schweitzer is firmly committed to ignoring the implications of her work, that these bones are very, very young. That secures her professional position. If she were to acknowledge this implication, that would be the end of her professional career. Eventually well-orchestrated condemnation would cost the author her job, as documented repeatedly in Ben Stein's movie *Expelled*.[55]

No Great Ape Fossils

"No fossils of any of the great apes -- gorillas, chimpanzees or orangutans -- have ever been found. As far as the fossil record is concerned, they never existed; and yet we know from the evidence of our own eyes that they did, and do."[56]

Though hundreds, perhaps thousands, of news releases, newspaper articles, magazine articles, radio, video, and internet announcements have proclaimed the discovery of a fossil of a great ape, none has ever been verified. Such discoveries have either proven to not be fossilized, and therefore much younger, or not to be a great ape. Some who carefully examine the evidence come to the reasoned conclusion that evolution is not scientifically valid. They do not find reproductive links in the evolutionary chain. Creationists are frequently charged with ignoring the evidence. Richard Dawkins claims to refute the fossil gap arguments in his numerous books, although he doesn't really refute them at all. He just dismisses the arguments with a bald, unsupported statement that large numbers of intermediates exist.

"Creationists are deeply enamored of the fossil record ...[They] repeat, over and over, the mantra that it is full of 'gaps': 'Show me your "intermediates!"' ...We [have]...massive numbers ... to document evolutionary history ... beautiful 'intermediates.' ... The fossil evidence for evolution in many major animal groups is wonderfully strong. Nevertheless there are, of course, gaps, and creationists love them obsessively."[57]

The truth is that there is nothing for mankind to link to. Neanderthals interbred with modern man. A skull found in *Pestera cu Oase,* "The Cave of Bones" in Romania, has characteristics of both so-called "species."

"The skull bearing both older and modern characteristics is discussed in a paper by Erik Trinkaus of Washington University in St. Louis. The report appears in today's [January 15, 2007] issue of *Proceedings of the National Academy of Sciences*.

"...The researchers said the skull had the same proportions as a modern human head and lacked the large brow ridge commonly associated with

Neanderthals. However, there were also features that are unusual in modern humans, such as frontal flattening, a fairly large bone behind the ear and exceptionally large upper molars, which are seen among Neanderthals and other early hominids.

“Such differences raise important questions about the evolutionary history of modern humans,” said co-author João Zilhão of the University of Bristol, England.[58]

Therefore, Neanderthals are men. Without any fossil evidence of creatures for man to evolve from, evolution of humans is a religious leap of faith. There is no evidence of any link and no fossil evidence of anything to link to.

Mitochondrial DNA

Any link would controlled by DNA. The most important aspect of the soft tissue discovery in the dinosaur is the DNA. DNA is in all cells of all living organisms. The tiniest cell in the human body contains the same DNA in the nucleus as the largest cell. This genetic coding determines hair color, shape of our nostrils, and every other detail of our body. As every student of high school biology knows, this DNA inside the nucleus contains genetic material from both our mother and our father. But many people are unaware of another type of DNA known as mitochondrial DNA. Mitochondrial DNA exists within the cell’s mitochondria outside of the nucleus and contains only genetic material from our mothers. Since there is no mixing of genetic material, a child’s mitochondrial DNA should be an exact duplicate of the mother’s mitochondrial DNA.

Mitochondrial DNA controls individual cell functions, not genetically encoded information. Infinite generations with genetically different fathers can have identical mitochondrial DNA but radically different physical characteristics. This should allow for a trace of mitochondrial DNA back to the beginning the human race. We could perform this trace, except for the problem of genetic mutations. Since a child's normal mitochondrial DNA is identical to his mother's, any mutation starts a new line. Minor mutations are difficult to detect and major mutations, such as those caused by massive radiation exposure, produce children who are unable to survive. But a few extremely rare mutations have produced traceable lines.

In 1997, *Nature Genetics* published an article with the innocuous title "A high observed substitution rate in the human mitochondrial DNA control rate." An abstract is available online for free. Though the technical language is absolutely necessary for accuracy, it distracts the average reader. Even with nothing but the abstract, the force of this study is overwhelming.

"...We report a direct measurement of the intergenerational substitution rate in the human CR [Control Region – a section of the DNA that controls other sections of DNA]. We compared DNA sequences of two CR hypervariable segments from close maternal relatives, from 134 independent mtDNA lineages spanning 327 generational events. Ten substitutions were observed, resulting in an empirical rate of 1/33 generations, or 2.5/site/Myr. This is roughly twenty-fold higher than estimates derived from

phylogenetic analyses. This disparity cannot be accounted for simply by substitutions at mutational hot spots, suggesting additional factors that produce the discrepancy between very near-term and long-term apparent rates of sequence divergence. The data also indicate that extremely rapid segregation of CR sequence variants between generations is common in humans, with a very small mtDNA bottleneck."[59]

What this study did was take mitochondrial DNA samples from living volunteers and compared it to mitochondrial DNA from their ancestors. The test subjects had to have accurate records of their ancestors for hundreds of years, so all test subjects had to be European. Corpses had to be available with accurate records and preserved well enough that mitochondrial DNA could be obtained. The study went back approximately 550 years.

The test results proved that mutations happen at a significantly higher rate than geneticists thought possible. The resulting statistical analysis came to two stunning conclusions.

First, applying this data to other studies gives an approximate date of 6,000-6,500 BP (Before Present) for Mitochondrial Eve.[60]

Catastrophes powerful enough to move marine ammonites to the top of Mount Everest, and tilt Lake Titicaca, could easily, temporarily and dramatically, increase the mutation rate, making even the 6,000-year-old date too old.

Second is the "very small mtDNA bottleneck." This means that at some point in the history of the human race there were very few women. Bryan

Sykes, author of *The Seven Daughters of Eve* believes that all Europeans currently alive can be traced back to one of seven women. Though Sykes is a well-respected English geneticist, as a committed Secular Humanist he makes two common though critical errors. These two errors have become the "standard" which any geneticist who expects to be published must adhere to. These errors are required by the religious dogma of Secular Humanism. Because his religious dogma demands that the human race is much older than the scientific evidence shows, he makes the leap of faith that the 2.5/site/Myr was lower in the past. His second error is assuming by a leap of faith, without any scientific evidence, that these seven women represent the smallest point of the bottleneck.

The Bible says that the sons of Noah brought their wives on the ark. These three women are the smallest point of the bottleneck. Their daughters or granddaughters might be the seven women of Bryan Sykes. Since conditions at that time could easily cause increased genetic mutations, their daughters, granddaughters or great granddaughters could easily be the seven women isolated by Bryan Sykes. When you read *The Seven Daughters of Eve,* understand that the last part of the book is devoted to fictional indoctrination. He novelizes each of these women and places each of them in a "cavewoman" setting. This is as antiscientific an approach as possible. Nevertheless the book was published. The scientific evidence of Mitochondrial DNA concludes that the human race is much younger than the published dates, less than 10,000 years, and that at one time in the very recent past the entire human race had no more than seven women. Bryan Sykes added the cavewoman fictions to assure publication. Such fabrications are the essence of the academic requirements for upholding the secularist dogma.[56]

He also claims that this genetic bottleneck of seven women is only for women of European descent, not the entire human race.

A much better study, without the approval of the academic community, therefore without the publicity, is Dr. Robert W. Carter's *Adam, Eve and Noah vs Modern Genetics*. This article also explores why the Y chromosome supports a recent genetic bottleneck. Dr. Carter states, "There are three main mitochondrial DNA lineages found across the world. The evolutionists have labeled these lines 'M', 'N', and 'R'..."[61]

Thermodynamics

Engineers can earn a PhD in many subcategories of the massive field of thermodynamics. The dynamics of heat affects all other aspects of engineering. The first two laws of thermodynamics are well-tested and well-proved. While the implications are massive, the first two laws are simple. The first law of thermodynamics: The conservation of energy. Energy can be changed from one form to another, but it cannot be created or destroyed. The total amount of energy and matter in the universe remains constant, merely changing from one form to another.

Albert Einstein popularized the first law of thermodynamics with his famous proposition "If a body at rest emits a total energy of E while remaining at rest, then the mass of that body decreases by E/c^2."

This is oversimplified into $E=mc^2$. Though technically this only applies to a body at rest, it is a workable understanding of the concept that matter and energy are interchangeable. In this formula, "c" represents the speed of light, in a vacuum, 299, 792, 458 meters per second or 186,300 miles per second.

Here is a link to a blog which works through the Lorentz transformation to explain how Albert Einstein arrived at his final formula. [62]

According to the first law of thermodynamics the universe is either 1) eternal, 2) created out of nothing, or 3) came into existence through some process that we know nothing about.

The Second Law of Thermodynamics states "in all energy exchanges, if no energy enters or leaves the system, the potential energy of the state will always be less than that of the initial state." This is also commonly referred to as entropy. Another way of stating entropy is that all energy changes are in a downward direction.[63] Without a belief in the religious dogma that the universe is infinite and can therefore obtain energy from "outside" the system (which is scientifically unprovable), entropy will eventually lead to a "heat death" for the entire universe. Everything in the universe will be motionless, at a uniform temperature near absolute zero, broken down into subatomic particles.

Each one of these examples of physical evidence has the same message. To repeat a quote from Dr. Danny R. Faulkner at the beginning of this chapter:

"While the early faint Sun paradox does not tell us that the Solar System is only thousands of years old, it does seem to rule out the age being billions of years."[64]

We can rephrase this statement to make it more inclusive. While none of these points in this chapter tell us that the earth is only thousands of years old, each one individually and all of them collectively seem to rule out the age being billions of years.

We close with the words of the committed Secular Humanist, Isaac Asimov, as he condemns not only his own belief system, but also the belief systems of everyone who reject the evidence for the Word of God.

> *"I believe in evidence. I believe in observation, measurement, and reasoning, confirmed by independent observers. I'll believe anything, no matter how wild and ridiculous, if there is evidence for it. The wilder and more ridiculous something is, however, the firmer and more solid the evidence will have to be."*[65]
>
> Isaac Asimov

---

1 Charles Darwin, "Letter to Asa Gray," (Harvard Professor of Biology), 18 June, 1857.

2 John Adams, "Argument in defence of the [English] soldiers in the Boston Massacre trial," December 1770.

3 Mark Twain, *Life on the Mississippi,* first edition published by Osgood and Company, 1883.

4 Samuel Johnson, *The History of Rasselas, Prince of Abissinia,* 1759.

5 Dennis Prager, "Breastfeeding as a Religion," World Net Daily, http://wnd.com/, posted November 11, 2003 1:00 am Eastern.

6 Richard W. Pogge, Astronomy 162: Introduction to Stars, Galaxies, & the Universe, 2006, http://www.astronomy.ohio-state.edu/-pogge/Ast162/Unit2/sunshine.html

7 "Fusion", *Nobelprize.org* updated 9 Mar 2013 http://nobelprize.org/nobel_prizes/physics/articles/fusion/sun_4.html

8 Ian T Durham, "Hans Bethe" (Biographic article about Hans Bethe), Saint Anselm College, Goffstown, NH, http://www-history.mcs.st-and.ac.uk/Biographies/Bethe.html

9 Akridge, R. 1980. "The Sun Is Shrinking." *Acts & Facts*. 9 (4) and Faulkner, D. 1998. "The Young Faint Sun Paradox and the Age of the Solar System." *Acts & Facts*. 27 (6). [these two articles are quoted and paraphrased from throughout this section]

10 Roger Freedman, Robert Geller, William J. Kaufmann, *Universe: The Solar System,* Macmillan, New York, NY, 2010.

11 Two links to articles which examine the early faint sun paradox problem in more detail:

http://creation.com/our-steady-sun-a-problem-for-billions-of-years

http://creation.com/the-young-faint-sun-paradox-and-the-age-of-the-solar-system

12 Daily Mail online, October 3, 2011 http://www.dailymail.co.uk/sciencetech/article-2044480/NASAs-SDO-satellite-shows-boiling-sun-stunning-detail.html#ixzz2MnMZZ5lS

13 Ethan Siegal PhD in theoretical astrophysics at the University of Florida *Starts with a Bang*. "How the Sun works, from the inside out", originally posted August 12, 2011.

14 D. Russell Humphreys, Steven A. Austin, John R. Baumgardner, and Andrew A. Snelling, "Helium Diffusion Age of 6,000 Years Supports Accelerated Nuclear Decay," *Creation Research Society Quarterly Journal. (CRSQ)* Vol 41 No 1 June 2004, *Creation Research.org,* Copyright © 2004 by Creation Research Society. Also Dr. Don DeYoung, *Thousands, Not Billions: Challenging an Icon of Evolution Questioning the Age of the Earth,* Green Forest, AR: Master Books, Inc., 2005.

15 Utah Geological Survey, *Utah.gov*.

16 Radiometric Dating background from the position of those who believe it to be valid can be studied on *Talk Origins.org.,* especially the article *Radiometric Dating and the Geological Time Scale: Circular Reasoning or Reliable Tools?* Andrew MacRae Copyright 1997-2004 [Text last updated: October 2, 1998] and also these two books: G. Brent Dalrymple, *The Age of the Earth*. Stanford University Press: Stanford, CA, 1991. G. Faure, *Principles of Isotope Geology,* 2nd. edition. John Wiley and Sons: New York, NY, 1986.

17 John Michael Fischer, *Dinosaur bones have been Carbon-14 dated to less than 40,000 years,*

http://newgeology.us/presentation48.html, 2012-2014.

18 Press release, Public Information Office, Jet Propulsion Laboratory, California Institute of Technology, *NASA,* July 21, 1994.

19 Andrew A. Snelling, "The Earth's magnetic field and the age of the Earth," first published: *Creation (Creation Ministries International),* 13(4):44-48 September 1991.

20 Humphreys, *Earth's Magnetic Field is Decaying Steadily–with a Little Rhythm,* CRSQ http://www.creationresearch.org/crsq/articles/47/47_3/CRSQ%20Winter%202011%20Humphreys.pdf

21 Jonathan Sarfati, *The Earth's Magnetic Field: Evidence That the Earth Is Young* http://creation.com/the-earths-magnetic-field-evidence-that-the-earth-is-young

22 Alexandra Witze, "Geomagnetic Flip-Flops in a Flash," *Science News,* September 25, 210 Vol. 178, Number 7, and R.S. Coe, and M. Prevot, 1989. "Evidence suggesting extremely rapid field variation during a geomagnetic reversal," *Earth and Planetary Science Letters, Elsevier*, Amsterdam, Netherlands, vol. 92, pp. 296-297.

23 http://www.answersingenesis.org/home/area/cfol/ch3-grand-canyon.asp

24 Type/Process: Pyroclastic Flow Volcanic Status: Historical Image Number: 029-008 Photographer: Norm Banks, 1980 (U.S. Geological

Survey)
Summit Elevation: 2549 meters
Latitude/Longitude: 46.20 N / 122.18 W
Timeframe: Last known eruption 1964 or later
Region: Canada and Western USA

25 G.R. Morton, (An old-earth supporter who calls himself a creationist apologist) "Young-Earth Arguments: A Second Look," 1998, *home.entouch.net.*

26 J.B. Delair and E.F. Oppe, "The Lost Sea of Andes," in Charles Hapgood's *The Path of the Pole,* Chilton Book Company, Philadelphia, PA, 1970.

27 Donald W. Patten and Samuel R. Windsor, "Catastrophic Theory of Mountain Uplifts (A Crustal Deformation Theory)," *Catastrophism and Ancient History* Vol. XIII Part 1 January 1991.

28 "Ancient temple found under Lake Titicaca," *BBC News, UK*, Wednesday, 23 August, 2000, 11:04 GMT 12:04.

29 J. P. Davidson, W. E. Reed, and P. M. Davis, "The Rise and Fall of Mountain Ranges," in *Exploring Earth: An Introduction to Physical Geology,* Upper Saddle River, New Jersey, Prentice Hall, 1997.

30 Erich A. Von Fange, "Time Upside Down," *Creation Research Quarterly,* June 1974.

31 *National Geographic*, July 26, 2008, "Tiny Fossils reveal Warm Antarctic Past," and *AntarcticConnection.com* reports on the Antarctic research stations.

32 Sean Pitman, M.D., in a PowerPoint presentation titled "Ancient Ice," created in Jan 2006, including

testimony from a phone interview with Bob Cardin, project manager to recover one of the P38s lost on the glacier.

33 Fred Hall, "Ice Cores Not All That Simple," *AEON II:* 1, 1989:199.

34 "Superbridge," *NOVA*, PBS, November 12, 1997.

35 Dr. Nathan Green, online course overview for GEO.101, "Introduction to Geology," Spring 2006, University of Alabama.

36 Jonathan Sarfati, *Salty Seas: Evidence for a Young Earth, Creation* 21(1):16–17, December 1998, http://creation.com/salty-seas-evidence-for-a-young-earth

37 *GeorgiaEncyclopedia.org* gives background on the canyon but attributes its underlying geology to the "millions of years" formation theory. See also "Canyon Creation," by Rebecca Gibson, *Answers in Genesis,* September 2000.

38 National Park Service report on Wall Arch collapse with before and after photos, August 4-5, 2008.

39 Photo of recently-formed stalactites courtesy of Creation Ministries International (see footnote 44 for commentary on the Australian Mine photo). Lincoln Memorial Photo from National Parks services website.

40. National Park Services Website.

41 Dave E. Matson, "How Good Are Those Young-Earth Arguments?" on *Infidels.org*. copyright 1995.

42 *GORP.com* (Great Outdoor Recreation Page.)

43 "Cave Reveals Southwest's Abrupt Climate Swings During Ice Age," *Science Daily.com,* January 25, 2010.

44 http://creation.com/stalactites-do-not-take-millions-of-years

45 American Museum of Natural History website, "How Old are Kimberlites and Diamonds?" see also *Novori.com* for history and manufacture of laboratory produced diamonds.

46 Andrew Snelling (Dr.), "Radiocarbon in Diamonds Confirmed," *Answers in Genesis,* November 7, 2007. (This study was conducted during the *RATE (Radioisotopes and the Age of The Earth)* research project at the Institute for Creation Research.)

47 "Coal, Volcanism and Noah's Flood," *TJ (Technical Journal)* 1(1): *Creation Ministries International*, 11–29 April 1984.

48 A.A. Snelling, "The Recent Origin of Bass Strait Oil and Gas," Creation, 5 (2):43–46 March 1982.5

49 Phil McCafferty, "Instant petrified wood?" *Popular Science,* October 1992, pp. 56-57. also Hamilton Hicks, 'Mineralized sodium silicate solutions for artificial petrification of wood,' United States Patent Number 4,612,050, September 16,1986, pp. 1-3. As cited by: Steven Austin, *CatastroRef—*"Catastrophe Reference Database: Catastrophes in Earth History, Geologic Evidence, Speculation and Theory," *Institute for Creation Research,* San Diego. Entry no. 267.

50 *Petrified Wood: Fast or Slow?* http://creation.com/petrified-wood-fast-or-slow

51 Michael Oard, *Frozen in Time: The Wooly Mammoth, The Ice Age and the Bible,* Green Forest, AR: Master Books, Inc., 2004.

52 http://creation.com/dino-dna-bone-cells#endRef5

53 http://www.nature.com/news/can-a-mammoth-carcass-really-preserve-flowing-blood-and-possibly-live-cells-1.13103

54 Schweitzer, Mary H. and Jennifer L. Wittmeyer, North Carolina State University; John R. Horner, Montana State University; Jan B. Toporski, Carnegie Institution of Washington Geophysical Laboratory. “Soft-Tissue Vessels and Cellular Preservation in Tyrannosaurus Rex.” *Science,* March 25, 2005. (NC State, the N.C. Museum of Natural Sciences and the National Science Foundation funded the research.)

55 Kevin Miller, Ben Stein, writers, Producers Logan Craft, Walt Ruloff and John Sullivan, Director Nathan Frankowski. Assoc. Prod. Mark Mathis. Ed. Simon Tondeur. *Expelled: No Intelligence Allowed.* © 2008 Premise Media Corporation, Rampart Films Production.

56 Bryan Sykes, *The Seven Daughters of Eve: The Science That Reveals Our Genetic Ancestry.* W.W. Norton, New York, N.Y., 2001.

57 Richard Dawkins, *The Greatest Show on Earth,* Free Press, Simon and Schuster, New York, NY, also by Bantam Press Transworld Publishers in Great Britain, 2009.

58 Randolph E. Schmid (Associated Press), "Skull Suggests Interbreeding of Neanderthal and Modern Man," *The Denver Post,* January 15, 2007.

59 Parsons, Thomas J., et. al. "A high observed substitution rate in the human mitochondrial DNA control region." Nature Genetics 15, 363 - 368 (1997).

60 http://creation.com/refuting-evolution-chapter-6-humans-images-of-god-or-advanced-apes#r25

61 http://creation.com/noah-and-genetics

62 http://terrytao.wordpress.com/2007/12/28/einsteins-derivation-of-emc2/

63 M.J. Farabee, *The Online Biology Book* (Farabee is a member of the Biology faculty at Estrella Mountain Community College, Avondale, Arizona. *emc.maricopa.edu*

64 Faulkner, D. 1998. "The Young Faint Sun Paradox and the Age of the Solar System." Acts & Facts. 27 (6).

65 Isaac Asimov, *The Roving Mind.* Prometheus Books, 1997.

5. What Is the Scientific Evidence for Intelligent Design?

The simplest explanation or strategy tends to be the best one.[1]
Ockham's Razor

Truth is ever to be found in simplicity, and not in the multiplicity and confusion of things.[2]
Sir Isaac Newton

Mathematics is the language with which God has written the universe.[3]
Galileo

The 2008 Ben Stein movie *Expelled: No Intelligence Allowed* documents the intolerance of the academic community toward anyone who believes in any form of Intelligent Design. Intelligent Design is simply the observation that the universe and everything in it are designed. It is a scientific fact that all known properly functioning systems of any type had a designer. The material universe shows evidence of design. Therefore, there must be a designer. Most rational cultures, modern as well as ancient, understand the scientific fact of Intelligent Design. The most simple, basic one-celled organism is vastly more complicated than the entire Internet. It is probably not a stretch to say that the complexity of the combined total of everything mankind has ever created is less complicated than a single-celled organism. Every religion that has ever existed understands this.

Every religion, that is, except the Established Religion of Secular Humanism. What is called Intelligent Design today is not Christianity. Hinduism believes in Intelligent Design. Hinduism also believes that the universe is billions of years old, if not eternal. Hinduism believes that the universe is the product of thousands of gods and is cyclical. Hinduism has little, if anything, in common with Christianity except for its recognition of the scientific fact that the universe is designed and not a random accident. Many of America's Founding Fathers were deists. They believed that God created the universe, and then left it to run according to natural laws. In that sense, they have more in common with Hinduism than Christianity. Pantheists, pagans, witches, Druids, Christians, Taoists, Buddhists, Hindus, Mormons, Muslims, Shamanists, Deists, Shintoists all recognize the scientific fact that the universe is designed.

The movements of stars, planets and moons are so regular that we can set our clocks and calendars by them. We can also navigate by them. Atomic particles vibrate with such regularity that atomic clocks are the most accurate timepieces known to man. Chemical reactions are so consistent that we can base technology on that consistency. Arithmetic, geometry, trigonometry, algebra, calculus and all other forms of mathematics are consistent enough to be the foundation of engineering, accounting and architecture. And yet the Established Religion of Secular Humanism will not only fire, but blacklist anyone who so much as mentions in a single sentence that these processes came into existence in any way other than by random chance.

The Established Religion of Secular Humanism censors all communication which might give information about intelligent design. They also substitute propaganda for information with the deliberate intention of misleading people. A reasonable person would not believe that a car, an airplane or even a paperclip was the product of random chance. Why then the fanatical belief that life itself is the product of random chance?

Louis Pasteur was a champion of the concept that life begets life. He opposed the doctrine of spontaneous generation before the concept of microbes was well known. To prove that there is no spontaneous generation, he heated milk. His heating process killed microbes. Since the microbes caused milk to spoil, the death of these microbes caused the milk to remain good for a longer period of time. Pasteurization is the scientific proof used in textbooks throughout the world that there is no spontaneous generation. If all life comes from previous generations, where did life originate?

This question is the foundation of mankind's most ancient documents, such as *The Epic of Gilgamesh, The Pyramid Texts* and the Bible. Many ancient religions taught that gods seeded life on earth. Modern followers of these same concepts use terms like ancient astronauts or aliens, but they mean the same thing. Life on earth began with material life somewhere else in the material universe. This line of reasoning only shoves the question of ultimate origins back farther in time without answering the basic questions; where did the material universe come from and where did life originate?

Physics provides us some vital information for answering these questions. The most important areas of study are Einstein's special and general theories of relativity and the first two laws of thermodynamics. Even without knowing exactly what it means, everyone in any industrialized country has seen the equation $e=mc^2$. In this equation, e stands for energy, m stands for mass and c stands for the speed of light. Written out, it says energy equals mass times the speed of light squared. It means that mass (which is matter or material) and energy are interchangeable. That is, matter can become energy and energy can become matter. It also means that as matter approaches the speed of light, there are important changes in gravity and time. But first we need to understand the major types of energy. The type of energy we are most familiar with is mechanical, such as a ball rolling downhill. The second is chemical, such as soap dissolving grease or gasoline burning. The third is electrical, such as energy stored in a battery. The fourth is thermonuclear, such as an atomic bomb. The changes between matter and chemical, mechanical and electrical energy are all around us every day. We eat food (matter). The food is changed chemically to provide us with mechanical energy (muscle movement), electrical energy (nerve impulses) or stored in our bodies as fat (energy converted back to material) to provide us with material which can be converted to chemical energy which will be converted to mechanical energy at a later time.

Chemical, mechanical and electrical energy, however, have immeasurably small effects on time. To have more than a tiny effect on time, we need gravity, and lots of it. The enormous mass of a star, such as our sun, has tremendous gravitational forces. Gravitational forces of stars can cause black holes, gravity wells, time distortions, event horizons and thermonuclear chain reactions to power them. These are not part of our everyday experiences, but they can have effects on time which are difficult to even imagine. Thanks to astronomy, we can witness these effects. All of this difficult to comprehend data still confirms the older yet still valid first two laws of thermodynamics.

The first two laws of thermodynamics have been tested. They are, in scientific terms, no longer just theory. They are laws. The first law simply states that matter can neither be created nor destroyed. Since matter and energy are simply different states of the same thing, then energy as well as matter can neither be created nor destroyed. Matter and energy simply change forms. The first law of thermodynamics was known, in some form, since the beginning of civilization. It forms the basis for many philosophies and religions which believe that the universe is eternal.

The second law of thermodynamics is less well known, but has been more thoroughly tested than the first law. It is also known as entropy. Simply put, entropy means that all energy transformations are in a downward direction. Any use of mechanical, chemical, electrical or thermonuclear energy results in some loss of available energy. For instance, the available chemical energy in one gallon of gasoline could propel an automobile hundreds of miles on level ground, if 100 percent of the potential chemical energy could be used to turn the drive tires. Whatever the actual number of miles is, there is an enormous energy loss in an automobile due to engine friction, incomplete combustion, heat loss (heat energy which is immediately lost and never converted into mechanical energy), and the inefficient use of the mechanical energy which is produced. Even the most efficient systems lose energy. The CPU of the computer used to create this document generates waste heat. Every process known to man uses more energy than it produces. Every energy transformation known to man illustrates entropy. Even a plant converting light, water and other chemicals into sugar and plant cells through photosynthesis requires far more energy than it produces. So natural processes as well as man-made ones are evidence for transformations in a downward direction.

Another way of stating the second law of thermodynamics is that the universe is gradually deteriorating from a position of greater complexity and order to an eventual end of complete disorder with only subatomic particles (perhaps hydrogen atoms) motion-less at a uniform temperature at or near absolute zero. Though we see things which are more complex than their basic material components, these are actually conservation forces. For instance, though the visible appearance of a plant seems more complex than the seed it grew from, the plant is only conserving the genetic information the seed already contained. The plant cannot be more complex than the genetic information it contains, which it got entirely from the seed. The seed, in its turn, simply conserves the genetic information it got from the plant. The only exceptions to this conservation are the loss of genetic information (mutations) and the artificial addition of genetic information by intelligence outside of the plant (genetic engineering).

The same principle of conservation holds true for the collective accomplishments of mankind. Men have always had the ability to work together to accomplish great things. Secularists say that man is evolving because he is learning through cooperation to produce greater and more complex things. But there are great structures and works of art from thousands of years ago that we cannot duplicate with today's technology. Today we use different technology, different methods. We build out of steel instead of out of stone. Still, every great construction project, every great work of art, every great engineering project, every great scientific idea; all of these and everything else that mankind does are simply conservation forces.

Even these conservation forces at work today require more energy than they produce. Where did the original matter and energy come from? And more important, where did the original order come from? Secular Humanists dogmatically assert that the order we see in the universe came from chaos and is becoming more orderly. But the law of entropy as proven by Lord Kelvin demonstrates that all energy transformations result in a loss of energy and that these changes are in a downward direction. The Bible teaches that the universe was created in perfect order and that sin is making the universe increasingly chaotic. Secularists believe that the original chaos is slowly evolving into a more orderly universe. Not only do they deny the first two laws of thermodynamics, but they disguise the deficiencies of their opposing theories by manufacturing immense periods of time.

These enormous ages for the Earth, the so-called "Deep Time," require a purely religious interpretation of the fossil record and a dogmatic belief in geologic ages. Deep Time is, however, the natural result of a preconceived rejection of an intelligent designer. Though the term "Deep Time" simple means that the earth is billions of years old, the details of Deep Time differ depending on the individual secularist. Practically no two agree on these questions of origins and ages. Many refuse to deal with the question of origin because their so-called scientific relativity is actually moral relativism. Since each individual secularist believes he has the right to have his own perspective on anything and everything, he has exalted himself to a god. To a Secularist, everything is relative. A Secularist believes that each person has the right to believe anything. Their only true and consistent point of agreement is that intelligent design proponents are wrong and Deep Time followed by spontaneous evolution is a fact.

As Richard Dawkins, the High Priest of Darwinism, said in 1976 in *The Selfish Gene,* "Today the theory of evolution is as much open to doubt as the theory that the earth goes round the sun, but the full implications of Darwin's revolution have yet to be widely realized."[4] He continues promoting his dogma with an attempt to shame anyone who would disagree. Secular Humanists frequently do this by invoking the favorite ally of those who believe life can emerge anywhere. “If superior creatures from space ever visit earth, the first question they will ask, in order to assess the level of our civilization is, ‘Have they discovered evolution yet?’”[4] Who can argue with a “superior creature?”

Dawkins goes on to say, “It is ironic that Ashley Montagu should criticize Lorenz as a ‘direct descendant of the “nature red in tooth and claw”’ thinkers of the nineteenth century...’ As I understood Lorenz’s view of evolution he would be very much at one with Montagu in rejecting Tennyson’s famous phrase. Unlike both of them, I think ‘nature red in tooth and claw’ sums up our modern understanding of natural selection admirably.”[4] Dawkins admits that there is considerable disagreement among Secular Humanists on how they present themselves and how they understand each other. It is certainly true that they are “red with tooth and claw”[5] from attacking any believer in intelligent design who dares to disagree with them.

In another of his works, *The Ancestor’s Tale, A Pilgrimage to the Dawn of Evolution,* Dawkins writes that, “Fossils are a bonus. A welcome bonus, to be sure, but not an essential one. It is worth remembering this when creationists go on (as they tediously do) about ‘gaps in the fossil record.’ The fossil record could be one big gap, and the evidence for evolution would still be overwhelmingly strong. At the same time, if we had only fossils and no other evidence, the fact of evolution would be overwhelmingly supported. As things stand, we are blessed with both.”[6] When you think of Dawkins, think of a fanatical "true believer" in Darwinism who believes what he believes and allows no facts to intrude.

Here is Dawkins on the subject of how Secular Humanists arrive at the millions of years of earth-age they depend on to make their processes work. "Age ranking depends on the assumption known as the Law of Superposition. For obvious reasons, younger strata lie atop older ones, unless the circumstances are exceptional." (In other words, they are this way unless they are some other way.) "Such exceptions, though they sometimes cause temporary puzzlement, are usually pretty obvious. A lump of old rock, complete with fossils, may be thrown on top of a younger stratum, say, by a glacier. The logic is complicated in practice though not in principle."[6] In the first line Dawkins uses the word "assumption." Without these assumptions, or preconceived ideas, or things you already believe, like the "fact" of evolution, you certainly will find the logic complicated. Why can't a proper disciple of Darwinism observe the facts and, if necessary, change their preconceptions?

Dawkins is a master of stating his position as if it were a simple, rational fact, the only position that could possibly be taken. So it is that no matter how absurd what he says sounds, the faithful must believe it or be heretics. In *The Extended Phenotype: The Long Reach of the Gene*, he makes this statement and dares anyone to disagree. "We have proved that there exists a trajectory of stepwise change connecting beetle to deer and, by implication, a similar trajectory from any modern animal to any other modern animal."[7] All of this absolute belief is essential to Secular Humanism, because doubters cannot be tolerated. Simply state a lie clearly and positively and make your followers believe it or make them and those who disagree ashamed of their stupidity.

No one seems more convinced of the power of positive assertion than Richard Lewontin. He is the Alexander Agassiz Research Professor at the Museum of Comparative Zoology, Harvard University. In spite of Richard Dawkins' assurances that evolution is confirmed, sensible, utterly impossible to question, Lewontin,"one of the foremost 'evolutionists,' Atheists and Materialists in the world,"[8] has been known to give vent to absolute certainties that are somewhat less than convincing at times. They certainly prove the point that secularists are true believers in their religion, even when they admit to its "patent absurdity."

> *"We take the side of science in spite of the patent absurdity of some of its constructs, in spite of its failure to fulfill many of its extravagant promises of health and life, in spite of the tolerance of the scientific community for unsubstantiated just-so stories, because we have a prior commitment, a commitment to materialism.*
>
> *"It is not that the methods and institutions of science somehow compel us to accept a material explanation of the phenomenal world, but, on the contrary, that we are forced by our a priori adherence to material causes to create an apparatus of investigation and a set of concepts that*

> *produce material explanations, no matter how counter-intuitive, no matter how mystifying to the uninitiated. Moreover, that materialism is an absolute, for we cannot allow a Divine Foot in the door.*"[9]

Lewontin also gave a lecture called "The Mind of a Toolmaker." He said that "scientists know next to nothing about how humans got so smart. 'We are missing the fossil record of human cognition, so we make up stories'."[10] Summarizing the rest of his lecture, UK science reporter James Randerson said,

> *"The eminent Harvard professor systematically dismissed every assumption about the evolution of human thought, reaching the conclusion that scientists are still completely in the dark about how natural selection prompted the massive hike in human brain size in the human line."*
>
> *"The main problem is the poor fossil record. Despite a handful of hominid fossils stretching back 4m [million—KB] years or so, we can't be sure that any of them are on the main ancestral line to us. Many or all of them could have been evolutionary side branches," Randerson continued, stating: "Worse, the fossils we do have are difficult to interpret. 'I don't have the faintest idea what the cranial capacity [of a fossil hominid] means,' Lewontin confessed. What does a particular brain size tell us about the capabilities of the animal attached to it?"*

> *[Lewontin] "is even skeptical that palaeoanthropologists can be sure which species walked upright and which dragged their knuckles. Upright posture is crucial for freeing up the hands to do other useful things."*

What, then, did Lewontin conclude regarding the prevailing status of ignorance that pervades the scientific community regarding the supposed evolution of humans? He said:

> *"We are in very serious difficulties in trying to reconstruct the evolution of cognition. I'm not even sure what we mean by the problem." Randerson concluded his summary of Lewontin's statements by observing: "All in all, despite thousands of scientific papers and countless National Geographic front covers, we have not made much progress in understanding how our most complicated and mysterious organ [brain—KB] came about."*[10]

Both Dawkins and Lewontin would assure us that though they don't have all the answers yet, and even if they have to "make up stories" to explain critical evolutionary concepts, their basic premise, evolution by chance over millions or billions of years, is unquestionably a fact. Dogmatically claiming great ages for the material universe, however, is extremely problematical because of the special and general theories of relativity. Einstein's theories of relativity, unlike the laws of thermodynamics, will likely always remain just theories because the number of variables is too great to test. Relativity does not prove the Bible or any of its positions. Nor does it disprove the Bible or any of its positions. It does, however, disprove Uniformitarianism. Relativity shows that there are too many variables to make dogmatic statements the way Secularists make them.

Since the most basic definition of relativity is that all measurements and observations are relative to the point of view of the observer, the number of possible observation points is critical. Although technically there are an infinite number, not all are possible or useful at a given point of view. Some intelligence must limit the number of possible points down to a finite number of workable combinations or no useful data can be obtained from observations. Something as basic as measuring temperature is relative. Measurement using degrees Kelvin is relative to absolute zero, where nothing moves, including subatomic particles. Measurement using degrees Celsius, or Centigrade, is relative to the freezing point and the boiling point of distilled water at sea level. Concerning Einstein's theory, narrow constraints and strict limitations must be applied even to make some aspect of it testable. To the extent that we can test it, not only is Einstein's theory correct, but also the number of variables affecting even one observer and one object observed are enormous.

A chess board has only 64 squares. Supercomputers fail to become chess grandmasters because there are more possible moves on a chessboard than there are molecules in the known universe. That doesn't seem possible but here is how it breaks down. On the very first move the first player can move a pawn either one or two spaces. There are eight pawns making a total of 16 possible first moves for the pawns. Each knight has two possible first moves and there are two knights for a total of four potential moves. That makes a total of twenty potential first moves. None of the other pieces can be moved on the first turn.

The second player has the same number of potential first moves. The potential first round moves consist of the twenty moves of the first player multiplied by the twenty moves of the second player for a total of 400 potential first turn moves. On the second turn you have the same twenty potential moves plus the potential moves created by the first moves: three moves for rook, seven for bishop, nine for queen, three for king, and eleven more for the knight. Knight, rook and bishop moves must be multiplied by two to account for the pairs of pieces. Seventy-six is the total of potential second moves and the second player has the same potential. Multiply the seventy-six potential moves of the first player by the four hundred potential moves from the first move of each player, then multiply that result by the seventy-six potential moves of the second player and you get a potential of 2,310,400 moves. Remember that at this point each player has only taken two moves. Each successive turn has more potential moves and the numbers increase dramatically.

Since the number of potential chess moves is so great, and the number of possible everyday decisions by a single human being is greater than the total number of chess moves in a game, then how is it possible to take into account everything that is relative to everything else in the universe? Men have the ability to make moral choices and to discard what is bad or evil. In chess, very few of the possible moves are good or useful. Accomplished players learn to concentrate on a few winning strategies and to ignore the vast number of non-winning moves. This limits the infinite number of possible moves down to a finite number of winning combinations.

As we learn to live life correctly, we must learn to concentrate on good decisions and ignore bad choices in life as we do in chess. Chess follows the scientific method in the sense that a player observes and records conditions, makes predictions, and modifies his play, possibly adopting a whole new strategy as the game progresses. Right decisions are based on correct observation, accurate interpretation, and proper adaptation to the conditions of the game. A chess player with rigid assumptions and unchangeable presuppositions will find it impossible to learn all the skills that make it possible to prevent bad moves before they are made.

Yet secularists approaching a scientific investigation have already eliminated certain possibilities. A chess player might, before even starting the game, have eliminated in his own mind the possibility of moving his pawns two spaces on their first move. He might say it is impossible to move his knights before moving any of his pawns. This is no different from what the secularist does to make his theory of deep time the only one possible. And if his opponent won't play by such absurd rules, he hurls insults, scatters the pieces, picks up his chessboard and goes home.

Though some limitations are necessary to make observations in physics and astrophysics possible, secularist uniformitarians concentrate on a vast number of untrue combinations and reject the obvious "winning strategies." The already huge number of possibilities in physics, especially astrophysics, jumps exponentially with the concepts of relativity and quantum physics. Secularists have no possibility of coming to correct conclusions. They mask these losing strategies by forbidding anyone to use good strategy.

Plato also attempted to illustrate the difficulty we have of understanding the universe. His *Republic* includes a parable about people who are from childhood chained to the wall of a cave, hearing only voices and seeing only shadows. Based on this "reality," they reward each other for "correctly" perceiving what the shadows carry, for "accurately" interpreting what the echoing voices say. They completely reject the testimony of those who were unchained, taken to the light and returned with more accurate information. Secularists, like the prisoners chained to a wall, are imprisoned by their own prejudice. They will deem any who disagree mad and destructive to their way of life and will try to stifle them. And they accuse theists of being the chained prisoners needing enlightenment.

---

1 Occam's (or Ockham's) razor is a principle attributed to the 14th century logician and Franciscan friar William of Ockham. The principle states that "Entities should not be multiplied unnecessarily." University of California, Riverside, *Math.ucr.edu*. Updated 1997 by Sugihara Hiroshi. Original by Phil Gibbs 1996.

2 Cited in *Rules for Methodizing the Apocalypse,* Rule 9, from a manuscript published in *The Religion of Isaac Newton* (1974) by Frank E. Manuel, p. 120, as quoted in *Socinianism And Arminianism : Antitrinitarians, Calvinists, And Cultural Exchange in Seventeenth-Century Europe* by Martin Mulsow, Jan Rohls, 2005.

3 As quoted in *Beginning Algebra* by Margaret L. Lial, Charles David Miller and E. John Hornsby, Harper Collins Collegiate Division, New York, NY, 1992.

4 Richard Dawkins, *The Selfish Gene.* London: Oxford University Press, 30th Edition, 2006.

5 (Author' Note: The quote "nature red in tooth and claw" comes from Alfred, Lord Tennyson's very long series of poems *In Memoriam A.H.H.* completed in 1849. Many evolutionists quote this phrase in support of their ideas of natural selection. When he began to write this poem, Tennyson questioned God's love and sovereignty over nature because of the death of a beloved friend. Parts of the poem comment on the pre-Darwinian writers who were beginning to promote man's reason and to shove God out of the Life Sciences. For an explanation of why Tennyson might not be the best person to quote on the subject if his conclusion in this poem is considered as part of the whole, see the Section Three Tennyson Appendix.)

6 Dawkins, *The Ancestor's Tale: A Pilgrimage to the Dawn of Evolution.* (Editorial research by Yan Wong) Boston, N.Y.: A Mariner Book, Houghton Mifflin, 2004.

7 Dawkins, *The Extended Phenotype: The Long Reach of the Gene.* London: Oxford University Press, 1982, 1999.

8 Kyle Butt, M.A, *"So We Make Up Stories" About Human Evolution,* Apologetics Press, 2008. (Butt also reviews and quotes from reporters Michael Balter and James Randerson in the articles cited below.)

9 Michael Balter, "How Human Intelligence Evolved—Is It Science or 'Paleofantasy'?" *Science* magazine. During the week of February 14-18, 2008, Dr. Lewontin was invited to speak at the American Association for the Advancement of Science's annual meeting held in Boston, Massachusetts and Balter reviewed his presentation in this article.

10 James Randerson, "We Know Nothing About Brain Evolution." *Guardian, UK.* Randerson, science correspondent, reported on a 2008 Lewontin speech titled, "Why We Know Nothing About the Evolution of Cognition." Randerson summarized some of Dr. Lewontin's statements.

Appendixes

Appendix One: Court Cases

Church of the Holy Trinity v. United States 1892 U.S. Supreme Court Church of the Holy Trinity v. United States, 143 U.S. 457 (1892) No. 143

Argued and submitted January 7, 1892 Decided February 29, 1892 143 U.S. 457

ERROR TO THE CIRCUIT COURT OF THE UNITED

STATES FOR THE SOUTHERN DISTRICT OF NEW YORK

Syllabus

The Act of February 26, 1880, "to prohibit the importation and migration of foreigners and aliens under contract or agreement to perform labor in the United States, its Territories, and the District of Columbia," 23 Stat. 332, c. 164, does not apply to a contract between an alien, residing out of the United States, and a religious society incorporated under the laws of a state, whereby he engages to remove to the United States and to enter into the service of the society as its rector or minister.

THE case is stated in the opinion.

MR. JUSTICE BREWER delivered the opinion of the Court.

Plaintiff in error is a corporation duly organized and incorporated as a religious society under the laws of the State of New York. E. Walpole Warren was, prior to September, 1887, an alien residing in England. In that month the plaintiff in error made a contract with him by which he was to remove to the City of New York and enter into its service as rector and pastor, and in pursuance of such contract, Warren did so remove and enter upon such service. It is claimed by the United States that this contract on the part of the plaintiff in error was forbidden by 23 Stat. 332, c. 164, and an action was commenced to recover the penalty prescribed by that act. The circuit court held that the contract was within the prohibition of the statute, and rendered judgment accordingly, 36 F.3d 3, and the single question presented for our determination is whether it erred in that conclusion.

The first section describes the act forbidden, and is in these words:

"Be it enacted by the Senate and House of Representatives of the United States of America in Congress assembled, that from and after the passage of this act it shall be unlawful for any person, company, partnership, or corporation, in any manner whatsoever, to prepay the transportation, or in any way assist or encourage the importation or migration, of any alien or aliens, any foreigner or foreigners, into the United States, its territories, or the District of Columbia under contract or agreement, parol or special, express or implied, made previous to the importation or migration of such alien or aliens, foreigner or foreigners, to perform labor or service of any kind in the United States, its territories, or the District of Columbia."

We find, therefore, that the title of the act, the evil which was intended to be remedied, the circumstances surrounding the appeal to Congress, the reports of the committee of each house, all concur in affirming that the intent of Congress was simply to stay the influx of this cheap unskilled labor.

But, beyond all these matters, no purpose of action against religion can be imputed to any legislation, state or national, because this is a religious people. This is historically true. From the discovery of this continent to the present hour, there is a single voice making this affirmation.

The commission to Christopher Columbus, prior to his sail westward, is from "Ferdinand and Isabella, by the grace of God, King and Queen of Castile," etc., and recites that "it is hoped that by God's assistance some of the continents and islands in the ocean will be discovered," etc. The first colonial grant, that made to Sir Walter Raleigh in 1584, was from "Elizabeth, by the grace of God, of England, Fraunce and Ireland, Queene, defender of the faith," etc., and the grant authorizing him to enact statutes of the government of the proposed colony provided that "they be not against the true Christian faith nowe professed in the Church of England." The first charter of Virginia, granted by King James I in 1606, after reciting the application of certain parties for a charter, commenced the grant in these words:

"We, greatly commending, and graciously accepting of, their Desires for the Furtherance of so noble a Work, which may, by the Providence of Almighty

God, hereafter tend to the Glory of his Divine Majesty, in propagating of Christian Religion to such People, as yet live in Darkness and miserable Ignorance of the true Knowledge and Worship of God, and may in time bring the Infidels and Savages, living in those parts, to human Civility, and to a settled and quiet government; DO, by these our Letters-Patents, graciously accept of, and agree to, their humble and well intended Desires."

Language of similar import may be found in the subsequent charters of that colony, from the same king, in 1609 and 1611, and the same is true of the various charters granted to the other colonies. In language more or less emphatic is the establishment of the Christian religion declared to be one of the purposes of the grant. The celebrated compact made by the pilgrims in the *Mayflower*, 1620, recites:

"Having undertaken for the Glory of God, and Advancement of the Christian Faith, and the Honour of our King and Country, a Voyage to plant the first Colony in the northern Parts of Virginia; Do by these Presents, solemnly and mutually, in the Presence of God and one another, covenant and combine ourselves together into a civil Body Politick, for our better Ordering and Preservation, and Furtherance of the Ends aforesaid."

The fundamental orders of Connecticut, under which a provisional government was instituted in 1638-39, commence with this declaration:

"Forasmuch as it hath pleased the Allmighty God by the wise disposition of his diuyne pruidence so to Order and dispose of things that we the Inhabitants and Residents of Windsor, Hartford, and Wethersfield are now cohabiting and dwelling in and vppon the River of Conectecotte and the Lands thereunto adioyneing; And well knowing where a people are gathered togather the word of God requires that to mayntayne the peace and vnion of such a people there should be an orderly and decent Gouerment established according to God, to order and dispose of the affayres of the people at all seasons as occation shall require; doe therefore assotiate and conioyne our selues to be as one Publike state or Comonwelth, and doe, for our selues and our Successors and such as shall be adioyned to vs att any tyme hereafter, enter into Combination and Confederation togather, to mayntayne and presearue the liberty and purity of the gospell of our Lord Jesus weh we now prfesse, as also the disciplyne of the Churches, weh according to the truth of the said gospell is now practiced amongst vs."

In the charter of privileges granted by William Penn to the province of Pennsylvania, in 1701, it is recited:

"Because no People can be truly happy, though under the greatest Enjoyment of Civil Liberties, if abridged of the Freedom of their Consciences, as to their Religious Profession and Worship; And Almighty God being the only Lord of Conscience, Father of Lights and Spirits, and the Author as well as Object of all divine Knowledge, Faith, and Worship, who only doth enlighten the Minds, and persuade and convince the Understandings of People, I do hereby grant and declare," etc.

Coming nearer to the present time, the declaration of independence recognizes the presence of the Divine in human affairs in these words:

"We hold these truths to be self-evident, that all men are created equal, that they are endowed by their Creator with certain unalienable Rights, that among these are Life, Liberty, and the pursuit of Happiness. . . . We therefore the Representatives of the United States of America, in General Congress, Assembled, appealing to the Supreme Judge of the world for the rectitude of our intentions, do, in the Name and by Authority of the good these Colonies, solemnly publish and declare," etc.; "And for the support of this Declaration, with a firm reliance on the Protection of Divine Providence, we mutually pledge to each other our Lives, our Fortunes, and our sacred Honor."

If we examine the constitutions of the various states, we find in them a constant recognition of religious obligations. Every Constitution of every one of the forty-four states contains language which, either directly or by clear implication, recognizes a profound reverence for religion, and an assumption that its influence in all human affairs is essential to the wellbeing of the community. This recognition may be in the preamble, such as is found in the Constitution of Illinois, 1870: "We, the people of the State of Illinois, grateful to Almighty God for the civil, political, and religious liberty which He hath so long permitted us to enjoy, and looking to Him for a blessing upon our endeavors to secure and transmit the same unimpaired to succeeding generations," etc.

It may be only in the familiar requisition that all officers shall take an oath closing with the declaration, "so help me God." It may be in clauses like that of the Constitution of Indiana, 1816, Art.

XI, section 4: "The manner of administering an oath or affirmation shall be such as is most consistent with the conscience of the deponent, and shall be esteemed the most solemn appeal to God." Or in provisions such as are found in Articles 36 and 37 of the declaration of rights of the Constitution of Maryland, 1867: "That, as it is the duty of every man to worship God in such manner as he thinks most acceptable to Him, all persons are equally entitled to protection in their religious liberty, wherefore no person ought, by any law, to be molested in his person or estate on account of his religious persuasion or profession, or for his religious practice, unless, under the color of religion, he shall disturb the good order, peace, or safety of the state, or shall infringe the laws of morality, or injure others in their natural, civil, or religious rights; nor ought any person to be compelled to frequent or maintain or contribute, unless on contract, to maintain any place of worship or any ministry; nor shall any person, otherwise competent, be deemed incompetent as a witness or juror on account of his religious belief, provided he believes in the existence of God, and that, under his dispensation, such person will be held morally accountable for his acts, and be rewarded or punished therefore, either in this world or the world to come. That no religious test ought ever to be required as a qualification for any office of profit or trust in this state, other than a declaration of belief in the existence of God; nor shall the legislature prescribe any other oath of office than the oath prescribed by this constitution."

Or like that in Articles 2 and 3 of part 1st of the Constitution of Massachusetts, 1780: "It is the right as well as the duty of all men in society publicly, and at stated seasons, to worship the Supreme Being, the great Creator and Preserver of the universe. . . . As the happiness of a people and the good order and preservation of civil government essentially depend upon piety, religion, and morality, and as these cannot be generally diffused through a community but by the institution of the public worship of God and of public instructions in piety, religion, and morality, therefore, to promote their happiness, and to secure the good order and preservation of their government, the people of this commonwealth have a right to invest their legislature with power to authorize and require, and the legislature shall, from time to time, authorize and require, the several towns, parishes, precincts, and other bodies politic or religious societies to make suitable provision at their own expense, for the institution of the public worship of God and for the support and maintenance of public Protestant teachers of piety, religion, and morality, in all cases where such provision shall not be made voluntarily."

Or, as in sections 5 and 14 of Article 7 of the Constitution of Mississippi, 1832: "No person who denies the being of a God, or a future state of rewards and punishments, shall hold any office in the civil department of this state. . . . Religion morality, and knowledge being necessary to good government, the preservation of liberty, and the happiness of mankind, schools, and the means of education, shall forever be encouraged in this state."

Or by Article 22 of the Constitution of Delaware, (1776), which required all officers, besides an oath of allegiance, to make and subscribe the following declaration: "I, A. B., do profess faith in God the

Father, and in Jesus Christ His only Son, and in the Holy Ghost, one God, blessed for evermore, and I do acknowledge the Holy Scriptures of the Old and New Testament to be given by divine inspiration."

Even the Constitution of the United States, which is supposed to have little touch upon the private life of the individual, contains in the First Amendment a declaration common to the constitutions of all the states, as follows: "Congress shall make no law respecting an establishment of religion, or prohibiting the free exercise thereof," etc., and also provides in Article I, Section 7, a provision common to many constitutions, that the executive shall have ten days (Sundays excepted) within which to determine whether he will approve or veto a bill.

There is no dissonance in these declarations. There is a universal language pervading them all, having one meaning. They affirm and reaffirm that this is a religious nation. These are not individual sayings, declarations of private persons. They are organic utterances. They speak the voice of the entire people. While, because of a general recognition of this truth, the question has seldom been presented to the courts, yet we find that in Updegraph v. Commonwealth, 11 S. & R. 394, 400, it was decided that "Christianity, general Christianity, is, and always has been, a part of the common law of Pennsylvania; . . . not Christianity with an established church and tithes and spiritual courts, but Christianity with liberty of conscience to all men."

And in People v. Ruggles, 8 Johns. 290, 294-295, Chancellor Kent, the great commentator on American law, speaking as Chief Justice of the

Supreme Court of New York, said: "The people of this state, in common with the people of this country, profess the general doctrines of Christianity as the rule of their faith and practice, and to scandalize the author of these doctrines is not only, in a religious point of view, extremely impious, but, even in respect to the obligations due to society, is a gross violation of decency and good order. . . . The free, equal, and undisturbed enjoyment of religious opinion, whatever it may be, and free and decent discussions on any religious subject, is granted and secured; but to revile, with malicious and blasphemous contempt, the religion professed by almost the whole community is an abuse of that right. Nor are we bound by any expressions in the Constitution, as some have strangely supposed, either not to punish at all, or to punish indiscriminately the like attacks upon the religion of Mahomet or of the Grand Lama, and for this plain reason, that the case assumes that we are a Christian people, and the morality of the country is deeply engrafted upon Christianity, and not upon the doctrines or worship of those impostors."

And in the famous case of Vidal v. Girard's Ex'rs, 2 How. 127, 198, this Court, while sustaining the will of Mr. Girard, with its provision for the creation of a college into which no minister should be permitted to enter, observed: "It is also said, and truly, that the Christian religion is a part of the common law of Pennsylvania."

If we pass beyond these matters to a view of American life, as expressed by its laws, its business, its customs, and its society, we find everywhere a clear recognition of the same truth. Among other matters, note the following: the form of oath universally prevailing, concluding with an appeal to the Almighty; the custom of opening sessions of all deliberative bodies and most conventions with prayer; the prefatory words of all wills,"In the name of God, amen;" the laws respecting the observance of the Sabbath, with the general cessation of all secular business, and the closing of courts, legislatures, and other similar public assemblies on that day; the churches and church organizations which abound in every city, town, and hamlet; the multitude of charitable organizations existing every where under Christian auspices; the gigantic missionary associations, with general support, and aiming to establish Christian missions in every quarter of the globe. These, and many other matters which might be noticed, add a volume of unofficial declarations to the mass of organic utterances that this is a Christian nation. In the face of all these, shall it be believed that a Congress of the United States intended to make it a misdemeanor for a church of this country to contract for the services of a Christian minister residing in another nation?

Suppose, in the Congress that passed this act, some member had offered a bill which in terms declared that if any Roman Catholic church in this country should contract with Cardinal Manning to come to this country and enter into its service as pastor and priest, or any Episcopal church should enter into a like contract with Canon Farrar, or any Baptist church should make similar arrangements with Rev. Mr. Spurgeon, or any Jewish synagogue with some eminent rabbi, such contract should be adjudged unlawful and void, and the church making it be subject to prosecution and punishment. Can it be believed that it would have received a minute of approving thought or a single vote? Yet it is contended that such was, in effect, the meaning of this statute. The construction invoked cannot be accepted as correct. It is a case where there was presented a definite evil, in view of which the legislature used general terms with the purpose of reaching all phases of that evil, and thereafter, unexpectedly, it is developed that the general language thus employed is broad enough to reach cases and acts which the whole history and life of the country affirm could not have been intentionally legislated against. It is the duty of the courts under those circumstances to say that, however broad the language of the statute may be, the act, although within the letter, is not within the intention of the legislature, and therefore cannot be within the statute.

The judgment will be reversed, and the case remanded for further proceedings in accordance with this opinion.

Epperson vs. Arkansas, 1968, United States Supreme Court

(Author's Note: Italicized material represents direct quotations. Material in regular type represents the author's comments.)

(The following paragraph is not part of a trial transcript, but is quoted from the website *Voices For Evolution*.)

In 1968, in Epperson v. Arkansas, the United States Supreme Court invalidated an Arkansas statute that prohibited the teaching of evolution. The Court held the statute unconstitutional on grounds that the

First Amendment to the U.S. Constitution does not permit a state to require that teaching and learning must be tailored to the principles or prohibitions of any particular religious sect or doctrine.

(Epperson v. Arkansas (1968) 393 U.S. 97, 37 U.S. Law Week 4017, 89S. Ct. 266, 21 L. Ed 228)

Following are excerpts from the Supreme Court transcript of this trial, taken from the website www.bc.edu/ bc_org/avp/cas/comm/ free_speech/epperson. Background material in the transcript explains that a teacher hired in 1964 to teach High School Biology completed her first year without incident. Her second year, however, a new textbook was obtained for her class which included a chapter on Darwin's theory. The teacher was apparently aware of the state's constitutional issue and decided to protect herself from disciplinary action before any was taken by suing the state to have the statue voided. The excerpt begins by stating the position of the attorney representing the State of Arkansas explaining how the state would interpret the statute.

On the other hand, counsel for the State, in oral argument in this Court, candidly stated that, despite the State Supreme Court's equivocation, Arkansas would interpret the statute "to mean that to make a student aware of the theory . . . just to teach that there was [103] such a theory" would be grounds for dismissal and for prosecution under the statute; and he said "that the Supreme Court of Arkansas' opinion should be interpreted in that manner." He said: "If Mrs. Epperson would tell her students that 'Here is Darwin's theory, that man ascended or descended from a lower form of being,' then I think she would be under this statute liable for prosecution."

In any event, we do not rest our decision upon the asserted vagueness of the statute. On either interpretation of its language, Arkansas' statute cannot stand. It is of no moment whether the law is deemed to prohibit mention of Darwin's theory, or to forbid any or all of the infinite varieties of communication embraced within the term "teaching." Under either interpretation, the law must be stricken because of its conflict with the constitutional prohibition of state laws respecting an establishment of religion or prohibiting the free exercise thereof. The overriding fact is that Arkansas' law selects from the body of knowledge a particular segment which it proscribes for the sole reason that it is deemed to conflict with a particular religious doctrine; that is, with a particular interpretation of the Book of Genesis by a particular religious group.

The following excerpt includes the specific opinion of Justice Black on this matter.

MR. JUSTICE BLACK, concurring.

I am by no means sure that this case presents a genuinely justiciable case or controversy. Although Arkansas Initiated Act No. 1, the statute alleged to be unconstitutional, was passed by the voters of Arkansas in 1928, we are informed that there has never been even a single attempt by the State to enforce it. And the pallid, unenthusiastic, even apologetic defense of the Act presented by the State in this Court indicates that the State would make no attempt to enforce the law should it remain on the books for the next century. Now, nearly 40 years after the law has slumbered on the books as though dead, a teacher alleging fear that the State might arouse from its lethargy and try to punish her has asked for a declaratory judgment holding the law unconstitutional. She was subsequently joined by a parent who alleged his interest in seeing that his two then school-age sons "be informed of all scientific theories and hypotheses ..."

Notwithstanding my own doubts as to whether the case presents a justiciable controversy, the Court brushes aside these doubts and leaps headlong into the middle of the very broad problems involved in federal intrusion into state powers to decide what subjects and schoolbooks it may wish to use in teaching state pupils. ... But, agreeing to consider this as a genuine case or controversy, I cannot agree to thrust the Federal Government's long arm the least bit further into state school curriculums than decision of this particular case requires. And the Court, in order to invalidate the Arkansas law as a violation of the First Amendment, has been compelled to give the State's law a broader meaning than the State Supreme Court was willing to give it. The Arkansas Supreme Court's opinion, in its entirety, stated that:

"Upon the principal issue, that of constitutionality, the court holds that Initiated Measure No. 1 of 1928, Ark. Stat. Ann. § 80-1627 and § 80-1628 (Repl. 1960), is a valid exercise of the state's power to specify the curriculum in its public schools. The court expresses no opinion on the question whether the Act prohibits any explanation of the theory of evolution or merely prohibits teaching that the theory is true; the answer not being necessary to a decision in the case, and the issue not having been raised."

It is plain that a state law prohibiting all teaching of human development or biology is constitutionally quite different from a law that compels a teacher to teach as true only one theory of a given doctrine. It would be difficult to make a First Amendment case out of a state law eliminating the subject of higher mathematics, or astronomy, or biology from its curriculum. And, for all the Supreme Court of Arkansas has said, this particular Act may prohibit that and nothing else. This Court, however, treats the Arkansas Act as though it made it a misdemeanor to teach or to use a book that teaches that evolution is true. But it is not for this Court to arrogate to itself the power to determine the scope of Arkansas statutes. Since the highest court of Arkansas has deliberately refused to give its statute that meaning, we should not presume to do so.

The Supreme Court struck down this Arkansas statue partly because it believed the state was violating the first and fourteenth amendments and partly because it believed the law to be too vaguely worded. The significant fact is in Black's statement that Arkansas had a chance to deal with the issue without federal interference and failed to do so. Black did not like even the idea that a Federal hand might be reaching into a state issue. Education specifics were supposed to be up to the states in those days. Arkansas gave away its state's right, and the rights of states in the future, by allowing the Supreme Court to establish this precedent.

The important issue of this case is just what Black said. The federal government made a ruling in a state issue and that shouldn't have happened. Time after time cases like this one have whittled away autonomy and replaced it with precedent. The power of the federal courts took away, little by little, the control of states over education and transferred it to the federal government. The Supreme Court has been a powerful tool in taking away our protection, our freedom, our rights and our property.

Segraves vs. State of California, 1981

(Author's Note: Italicized material represents direct quotations. Material in regular type represents the author's comments.)

(The following paragraph is not part of a trial transcript, but is quoted from the website Voices For Evolution)

In 1981, in Segraves v. State of California the Court found that the California State Board of Education's Science Framework, as written and as qualified by its anti-dogmatism policy, gave sufficient accommodation to the views of Segraves, contrary to his contention that class discussion of evolution prohibited his and his children's free exercise of religion. The anti-dogmatism policy provided that class discussions of origins should emphasize that scientific explanations focus on "how", not "ultimate cause," and that any speculative statements concerning origins, both in texts and in classes, should be presented conditionally, not dogmatically. The court's ruling also directed the Board of Education to widely disseminate the policy, which in 1989 was expanded to cover all areas of science, not just those concerning issues of origins.

(Segraves v. California (1981) Sacramento Superior Court #278978)

Following are excerpts from the oral presentation of the case. Superior Court Judge Irving Perluss adopted a very friendly and informal tone, insisting on the oral format, making light of the need for huge amounts of documentation, and so it's a bit difficult to find a clean copy of this transcript to draw from. Most of the sections are the opinion of the judge, but there is an instance where a witness, Dr. Meyer, is called upon to speak.

These excerpts come from a personal website built by Frank Fire, who says he is "a firefighter/ paramedic. (with a name like that, what else would I be?) I also built it to promote critical thinking and to promote the defense of the First Amendment." The text contained a number of typos and formatting errors which have been corrected for ease of reading, but the basic format is that of the original transcript.

The judge's tone seemed to imply that the case was not important enough to be recorded and everyone should just get together as friends and settle things. It established an important precedent, however, and Frank Fire performed an important service by posting a copy of the transcript.

And isn't it truly wonderful that in our country we can seek to invoke the awesome authority of the courts to assuage the feelings of a single child, a child. And in the final analysis, I believe that is what this case is all about. The play between establishment on the one hand, in terms of accommodation, and free exercise on the other. Now, fortunately -- I say "fortunately" because I'm the fellow that has to make the decision -- the issues have been narrowed here to the point where we are not faced with such a dilemma, and thus there is on contention here that evolution should not be taught in the public schools. I think you've heard me say on several occasions that if there were, it would be rejected as an impermissible accommodation, for that battle was fought and resolved by the Supreme Court of the United States, in Epperson versus Arkansas. Now, moreover, the Plaintiffs have disclaimed any interest in an accommodation which would require the teaching of special creation in the public schools. And I might say, in -- and of course, this is what they call, "dicta," this is not part of the decision in this case, but this is my -- my view -- that is was appropriate that they do so, for I have no doubt, whatever, that such an accommodation would be held to be violated with the establishment clause, and forbidden. I think this is so, as a matter of law. It was basically held to be such in the -- in the opinion of the California Attorney General in 58 Attorney General Opinions, 262. And of course, it was held in the decision of Daniels versus Waters in the Sixth Circuit, which was referred to during the course of our trial.

Now, the issue, simply stated, accordingly, is whether or not the free exercise of religion by Mr. Segraves and his children was thwarted by the instruction in science that children had received in school, and if so, has there been sufficient accommodation for their views?

The Court, in addition, is prepared to find and does find that the science framework, as written, and if qualified by the policy of the Board exemplified by

Exhibit N, does provide sufficient accommodation for the views of the Plaintiff. This is so, in my judgment, even if, as was alluded to by Mr. Turner, there is some problem about whether that was ever officially adopted as a policy by the Board because the fact is now that by virtue of the statement of the representative of the Board, more than one, not only the Attorney General but the representatives of the Board, that is current Board policy and shall remain as current Board policy until a Board changes it. I think all teachers -- I hope all teachers endeavor to follow the Code of Ethics and the administrative regulation that we have read 80130 of Title Five. But nevertheless, all of us conclude, and I conclude myself, sometimes are needed -- we need to be reminded of our responsibilities. It seems to me that what has happened here has developed from a lack of communication from the Board to the school to the classroom teacher. I think it is the emphasis on tolerance and understanding that should be communicated as a fundamental policy of the State Board of Education. This is true not only in science, but it's true throughout the entire public school system. I must add that it seems to the Court, also, that persons seeking tolerance and understanding must practice it, also. Only in this way can all of us enjoy the religious liberty which is our fundamental right.

Mr. Turner has already quoted from the concurring opinion of Justice Stewart, and Sherbert versus Vernor. I think it's worth repeating because those are resounding and beautiful words where he said, "I am convinced that no liberty is more essential to the continued vitality of the free society which our constitution guarantees than is the religious liberty protected by the free exercise clause explicit in the First Amendment and embedded in the Fourteenth." I think Justice Stewart has spoken well.

In the final analysis, ladies and gentlemen, counsel, all that Plaintiffs seek, in the Court's view, presently is contained in Board policy. It appears, however, that this Board policy may not have been communicated to all who should know of it, and who should be guided by that policy. As this is a Court of equity, it seems to the Court that an appropriate remedy may be fashioned.

It will be the order of the Court that there shall be disseminated to all the publishers, institutions, school districts, schools, and persons regularly receiving the science framework a copy of the Board policy set forth in Exhibit N. By this, the Court means, insofar as possible, the policies shall -- the policy shall be sent to those who have received the framework in the past. It shall be included in the framework disseminated in the future. It follows that if there are violations of this policy when disseminated it becomes a matter of concern for students and parents to adjust with their local teachers, their local schools, and their local school boards.

"Now, when you begin to think of textbooks that talk about belief, now to me belief is not a scientific word. One knows, one accumulates data, one has a comprehension of, one understands, one does a lot of things, but to me belief always, in my situation, has been something I associate with my theology. I would not like to see my theology and my science get mixed. I have never dealt with a scientific process where somebody says, 'I believe.' I have dealt with theological processes where one believes. In short, I think at that point you begin to mix epistemologies, and that's confusing."

And then I said to him,

"But I see, as I comprehend this case, we are talking about the very kind of disclaimer that you have just told us about, that you have just told us about, that at the beginning of a science textbook should there not be a statement --because not everyone is a scientist and knows all the background of scientists -- but shouldn't there be a statement saying, 'this does not deal with theology?'"

Dr. Mayer,

"Absolutely. I would -- I would say there should be a clear explanation that perhaps should run through the entire textbook within the student's mind what it is he's dealing with. He's dealing with science. We are not making a pretense to teach him music, art, poetry, theology, or any other discipline. Science does these things, and outside of that realm, science is not only moot, but might even be harmful."

Court,

"And, moreover, science is not dogmatic in that it is open ended and there is an absence of preset conclusions?"

The witness,

"Yes sir."

Now, there is one additional statement from Justice Stewart -- forgive me if I quote from him often, but our son clerked for him as a clerk, and so I -- I think he's a great man.

Justice Stewart also said in the concurring opinion, and in the Sherbert case, and these are the words that I felt were most pertinent to our case where he said, "And I think that the guarantee of religious liberty embodied in the free exercise clause affirmatively requires government to create an atmosphere of hospitality and accommodation to individual belief or disbelief. In short, I think our constitution commands the positive protection by government of religious freedom, not only for a minority, however small, not only for the majority, however large, but for each of us." I don't think any of us could really quarrel with that.

As I view this case, accordingly, counsel, I really don't believe that either side has lost. I truly believe that both sides have won. I think that we have all won because hopefully what we have achieved in this case is understanding.

California had a policy to make sure teachers didn't make evolution a dogma. The judge made a very kind and inclusive statement that everybody won but the man and his children lost the case.

The anti-dogmatism policy provided that class discussions of origins should emphasize that scientific explanations focus on "how," not "ultimate cause," and that any speculative statements concerning origins, both in texts and in classes, should be presented conditionally, not dogmatically.

This California statute may or may not have been sufficient to protect the religious beliefs of students. The intended result seems to have been that each side is free to believe what it will about origins. A consequence that may have been unintended is that by accepting and operating under this standard, by attempting to place both sides on an equal footing, Evolutionists reveal that they really do regard Evolution as a set of beliefs, in spite of the judge's contention that Science doesn't deal with beliefs.

McLean v. Arkansas Board of Education, 1982

(Author's Note: Italicized material represents direct quotations. Material in regular type represents the author's comments.)

(The following paragraph is not part of a trial transcript, but is quoted from the website Voices For Evolution.)

In 1982, in McLean v. Arkansas Board of Education, a federal court held that a "balanced treatment" statute violated the Establishment Clause of the U.S. Constitution. The Arkansas statute required public schools to give balanced treatment to "creation-science" and "evolution-science". In a decision that gave a detailed definition of the term "science," the court declared that "creation science" is not in fact a science. The court also found that the statute did not have a secular purpose, noting that the statute used language peculiar to creationist literature in emphasizing origins of life as an aspect of the theory of evolution. While the subject of life's origins is within the province of biology, the scientific community does not consider the subject as part of evolutionary theory, which assumes the existence of life and is directed to an explanation of how life evolved after it originated. The theory of evolution does not presuppose either the absence or the presence of a creator. (McLean v. Arkansas Board of

Education (1982) 529 F. Supp. 1255, 50 U.S. Law Week 2412)

Following are excerpts from the trial transcript. The source is *www.talkorigins.org/faqs/mclean-v-arkansas.*

Special attention should be paid to the notation in the preceding paragraph, however, the one that says While the subject of life's origins is within the province of biology, the scientific community does not consider the subject as part of evolutionary theory, which assumes the existence of life and is directed to an explanation of how life evolved after it originated. The theory of evolution does not presuppose either the absence or the presence of a creator. Keep this idea in mind for later consideration.

On March 19, 1981, the Governor of Arkansas signed into law Act 590 of 1981, entitled "Balanced Treatment for Creation-Science and Evolution-Science Act." The Act is codified as Ark. Stat. Ann. &80-1663, et seq., (1981 Supp.). Its essential mandate is stated in its first sentence: "Public schools within this State shall give balanced treatment to creation-science and to evolution-science." On May 27, 1981, this suit was filed challenging the constitutional validity of Act 590 on three distinct grounds.

First, it is contended that Act 590 constitutes an establishment of religion prohibited by the First Amendment to the Constitution, which is made applicable to the states by the Fourteenth Amendment. Second, the plaintiffs argue the Act violates a right to academic freedom which they say is guaranteed to students and teachers by the Free

Speech Clause of the First Amendment. Third, plaintiffs allege the Act is impermissibly vague and thereby violates the Due Process Clause of the Fourteenth Amendment.

The following excerpt is significant because it tells you that the courts have taken upon themselves the right to sole and indisputable power to tell us what the Constitution says and what it means. The courts can tell you how to think about this because people let them tell you how to think about it before. Now you're stuck.

There is no controversy over the legal standards under which the Establishment Clause portion of this case must be judged. The Supreme Court has on a number of occasions expounded on the meaning of the clause, and the pronouncements are clear. Often the issue has arisen in the context of public education, as it has here. In Everson v. Board of Education, 330 U.S. 1, 15-16 (1947), Justice Black stated:

The "establishment of religion" clause of the First Amendment means at least this: Neither a state nor the Federal Government can set up a church. Neither can pass laws which aid one religion, aid all religions, or prefer one religion over another. Neither can force nor influence a person to go to or to remain away from church against his will or force him to profess a belief or disbelief in any religion. No person can be punished for entertaining or professing religious beliefs or disbeliefs, for church-attendance or non-attendance. No tax, large or small, can be levied to support any religious activities or institutions, whatever they may be called, or what ever form they may adopt to teach or practice religion. Neither a state nor the Federal Government can, openly or secretly, participate in the affairs of any religious organizations or groups and vice versa. In the words of Jefferson, the clause ... was intended to erect "a wall of separation between church and State."

The Establishment Clause thus enshrines two central values: voluntarism and pluralism. And it is in the area of the public schools that these values must be guarded most vigilantly.

Designed to serve as perhaps the most powerful agency for promoting cohesion among a heterogeneous democratic people, the public school must keep scrupulously free from entanglement in the strife of sects. The preservation of the community from divisive conflicts, of Government from irreconcilable pressures by religious groups, or religion from censorship and coercion however subtly exercised, requires strict confinement of the State to instruction other than religious, leaving to the individual's church and home, indoctrination in the faith of his choice. [McCollum v. Board of Education, 333 U.S. 203, 216-217 (1948), (Opinion of Frankfurter, J., joined by Jackson, Burton, and Rutledge, J.J.)]

What Black said was proper and correct, up to the point where he inserted Jefferson's explanation about the "wall of separation." People have gotten to think that that's actually in the Constitution because of mentions like this, but it isn't. And it's been used to promote the idea that all mention of any religion must be excised from education because education must be a state function. What Thomas Jefferson actually meant is just the opposite of what Justice

Black says. Thomas Jefferson, the extreme leftist of his day, meant that the state must not interfere in the affairs of the Church; the wall was to protect the Church from the state. Thomas Jefferson meant what all of the founding fathers meant. No aspect of the federal government, including the courts, had any power, any authority to do anything at all with any aspect an establishment of religion. An Establishment of Religion is welfare, education and worship. Yet Justice Black in this case doesn't even stop there. He goes on to quote Frankfurter, and Frankfurter goes too far. He presupposes that the government must be the teacher or the people can't be cohesive.

Education wasn't originally a state function. It was usurped by the state from religious leaders and parents. In every society from ancient times to the present, where the government took over teaching the children, the seeds of that government's destruction were sown. The teaching of religion doesn't automatically mean that there will be strife or coercion or even indoctrination. Who puts up a fuss about religious instruction? People who don't want to be reminded that they are responsible to God. Look for them, and you'll see where the strife and coercion is coming from.

In the second place, no one even pretends that all religion has been excised from the public school. Yoga classes are taught in Physical Education. Teaching the Crusades in History class can't be done without bringing up religious issues, wrong ones and right ones. Teachers interweave culture and religion with Foreign Language study, and even studying Science brings up medical treatments at the Temple of Aesclipius and Egyptian pharmaceutical papyri ceremonially buried with priests.

The term “scientific creationism” first gained currency around 1965 following publication of *The Genesis Flood* in 1961 by Whitcomb and Morris. There is undoubtedly some connection between the appearance of the BSCS texts emphasizing evolutionary thought and efforts of Fundamentalist to attach the theory. (Mayer)

In the 1960’s and early 1970’s, several Fundamentalist organizations were formed to promote the idea that the Book of Genesis was supported by scientific data. The terms “creation science” and “scientific creationism” have been adopted by these Fundamentalists as descriptive of their study of creation and the origins of man. Perhaps the leading creationist organization is the Institute for Creation Research (ICR), which is affiliated with the Christian heritage College and supported by the Scott Memorial Baptist Church in San Diego, California. The ICR, through the Creation-Life Publishing Company, is the leading publisher of creation science material. other creation science organizations include the Creation Science Research Center (CSRC) of San Diego and the Bible Science Association of Minneapolis, Minnesota. In 1963, the Creation Research Society (CRS) was formed from a schism in the American Scientific Affiliation (ASA). It is an organization of literal Fundamentalists who have the equivalent of a master’s degree in some recognized area of science. A purpose of the organization is “to reach all people with the vital message of the scientific and historical truth about creation." Nelkin, The Science Textbook Controversies and the Politics of Equal Time, 66.

The preceding quote sets up the case being presented, the argument, not about whether teachers should have the freedom to teach what they believe is important and right, but whether a particular teaching must be attacked, discredited and beaten out of the schools. The court has already made itself the authority on what the state is and what religion is and how they can't be in the same room together. That includes the concept that the school is the state. A lengthy summary of the history of Fundamentalism is included in the transcript, because it's necessary for an inextricable bond to be formed between Creation Science and religious extremists. The transcript contends that proponents of this bill were on a religious crusade and that they attempted to conceal the fact to allow their bill to pass.

Senator James L. Holsted [was chosen to] introduce the act. Holsted, a self-described "born again" Christian Fundamentalist, introduced the act in the Arkansas Senate. He did not consult the State Department of Education, scientists, science educators or the Arkansas Attorney General. The Act was not referred to any Senate committee for hearing and was passed after only a few minutes' discussion on the Senate floor. In the House of Representatives, the bill was referred to the Education Committee which conducted a perfunctory fifteen minute hearing. No scientist testified at the hearing, nor was any representative from the State Department of Education called to testify.

The State failed to produce any evidence which would warrant an inference or conclusion that at any point in the process anyone considered the legitimate educational value of the Act. It was simply and purely an effort to introduce the Biblical version of creation into the public school curricula. The only inference which can be drawn from these circumstances is that the Act was passed with the specific purpose by the General Assembly of advancing religion. The Act therefore fails the first prong of the three-pronged test, that of secular legislative purpose, as articulated in Lemon v. Kurtzman, supra, and Stone v. Graham, supra.

An immediate attack is made on the bill's credentials. There were no scientists, no Department of Education input, no senate committee ruled on it, so it can't be valid. More about the issue of scientists later, but at least the last two entities can be identified as instruments of government control. The state must retain sole power here. A bill merely created and submitted by earnest, intelligent citizens is no good on its face.

A presentation of the basic tenets of Creation Science and Evolutionary Science follows in the transcript, setting the two in opposition to each other. An attack is made on this method of presentation as follows and the "true" definition of science is arrived at so that Creation Science can quickly be excluded entirely from serious consideration. The statement that "in a free society, knowledge does not require the imprimatur of legislation in order to become science" is almost laughable if it weren't such a lie. This whole case is about legislating what is and isn't Science.

In addition to the fallacious pedagogy of the two model approach, Section 4(a) lacks legitimate educational value because "creation-science" as defined in that section is simply not science. Several witnesses suggested definitions of science. A descriptive definition was said to be that science is what is "accepted by the scientific community" and is "what scientists do." The obvious implication of this description is that, in a free society, knowledge does not require the imprimatur of legislation in order to become science.

More precisely, the essential characteristics of science are:(1) It is guided by natural law;

(2) It has to be explanatory by reference to nature law;(3) It is testable against the empirical world;(4) Its conclusions are tentative, i.e. are not necessarily the final word; and (5) It is falsifiable. (Ruse and other science witnesses).

The term "Natural Law" comes from Plato and Aristotle. Aristotle acknowledged the possibility of an "unmoved mover," but Plato is famous for having thrown "the gods" (and the true God) out of his perfect society thousands of years ago. This court has taken an adversarial position carefully devised to exclude the possibility of Creation and set its own position up as the only possible fact. It's not fact, it's philosophy. The entire "definition" of Science presented here is Naturalism, a philosophy designed to exclude God, not true Science at all but a way to set up the worship of man's intellect as an extension of nature.

As many times as the transcript argues that Creation Science proofs don't prove anything, you would think that it would become clear that this premise as a definition of Science doesn't prove anything. About the only thing true about "Science" as defined here is that its conclusions certainly are tentative. Yet the position taken is that they are the final word. This is the dogma of a belief, a religion, not Science.

Note the folksy statements that Science is "what scientists do." Naturally that excludes Creationists, even though most of these proponents of Creation Science hold advanced degrees from accredited institutions in recognized scientific fields. Unless you presuppose that a Creationist can't be a scientist you would have to grant that these people really are scientists and therefore they also do what scientists do.

Creation science as described in Section 4(a) fails to meet these essential characteristics. First, the section revolves around 4(a)(1) which asserts a sudden creation "from nothing." Such a concept is not science because it depends upon a supernatural intervention which is not guided by natural law. It is not explanatory by reference to natural law, is not testable and is not falsifiable.

The same arguments are applicable to Evolution, or would be if Evolution were forced to justify its tenets by actually coming up with a viable theory to cover origins. Instead it is allowed to slide by with the explanation that it doesn't deal with origins. Actually, it does, but more along the lines of Eastern religions which allow for an endless cycle of time, progressively more billions of years. How can the statement that it presupposes the existence of life be satisfactory? It allows for all of Evolutionary theory to be based on presuppositions, and indeed that's all it is, suppositions based on the founding premise that we suppose there isn't any God who started everything.

Creation science as defined in Section 4(a), not only fails to follow the canons of dealing with scientific theory, it also fails to fit the more general descriptions of "what scientists think" and "what scientists do." The scientific community consists of individuals and groups, nationally and internationally, who work independently in such varied fields as biology, paleontology, geology, and astronomy. Their work is published and subject to review and testing by their peers. The journals for publication are both numerous and varied. There is, however, not one recognized scientific journal which has published an article espousing the creation science theory described in Section 4(a). Some of the State's witnesses suggested that the scientific community was "close-minded" on the subject of creationism and that explained the lack of acceptance of the creation science arguments. Yet no witness produced a scientific article for which publication has been refused. Perhaps some members of the scientific community are resistant to new ideas. It is, however, inconceivable that such a loose knit group of independent thinkers in all the varied fields of science could, or would, so effectively censor new scientific thought.

The reader is directed to Dr. Richard Sternberg's discussion of what happened to him when he published an article by Dr. Stephen Meyer which contained material related to Intelligent Design, not even espousing Creationism. (See the Section Three Appendix for specific examples of attacks on recognized members of the scientific community because of a connection with Intelligent Design.)

The trial transcript simply dismisses the possibility that the scientific community could be narrow minded. No doubt thousands of articles and books which have been refused publication could have been produced as evidence of this narrow-mindedness. The transcript later brings up the fact that a woman who was charged with developing curriculum materials was unable to find any that met her criteria for inclusion. In other words, her fruitless search proves one of two points, either that reputable scientists cannot get creationist material published or that her criteria is narrow-minded and exclusive of this material.

The defendants' argument would be more persuasive if, in fact, there were only two theories or idea about the origins of life and the world. That there are a number of theories was acknowledge by the State's witnesses, Dr. Wickramasinghe and Dr. Geisler. Dr. Wickramasinghe testified at length in support of a theory that life on earth was "seeded" by comets which delivered genetic material and perhaps organisms to the earth's surface from interstellar dust far outside the solar system. The "seeding" theory further hypothesizes that the earth remains under the continuing influence of genetic material from space which continues to affect life. While Wickramasinghe's theory about the origins of life on earth has not received general acceptance within the scientific community, he has, at least, used scientific methodology to produce a theory of origins which meets the essential characteristics of science.

It is a logical fallacy to create a definition and then make it the only definition anybody can use to prove anything. The preceding exposition of the theory of "genetic seeding" by Dr. Wickramasinghe is presented to prove that creation science isn't allowed to set itself off against Evolution because Evolution isn't the only theory of how things are the way they are. There's this one, which the Scientific community doesn't recognize and won't even fund further study about. Excellent choice. A theory with a completely unprovable, unexaminable and already generally discarded premise is at least good Science.

Robert Gentry's discovery of radioactive polonium haloes in granite and coalified woods is, perhaps, the most recent scientific work which the creationists use as argument for a "relatively recent inception" of the earth and a "worldwide flood." The existence of polonium haloes in granite and coalified wood is thought to be inconsistent with radiometric dating methods based upon constant radioactive decay rates. Mr. Gentry's findings were published almost ten years ago and have been the subject of some discussion in the scientific community. The discoveries have not, however, led to the formulation of any scientific hypothesis or theory which would explain a relatively recent inception of the earth or a worldwide flood. Gentry's discovery has been treated as a minor mystery which will eventually be explained. It may deserve further investigation, but the National Science Foundation has not deemed it to be of sufficient import to support further funding.

The National Science Foundation has decided not to fund this promising area of research for only one reason: It might disprove a few presuppositions Evolutionists cling to. If ever there was an example of narrow-mindedness in the scientific community, here it is. Demanding that Creationists must produce something "new" before they will get a hearing is a humorous thought coming from someone who resurrects Plato to craft a definition of Science. This is a common dismissal of Creationist evidence, that it's old so it isn't valid. How is age a worthy criteria for judging evidentiary value? If you can prove you were born in the United States thirty years ago is that evidence too old to be admitted as proof of citizenship?

In any event, if Act 590 is implemented, many teachers will be required to teach materials in support of creation science which they do not consider academically sound. Many teachers will simply forego teaching subjects which might trigger the "balanced treatment" aspects of Act 590 even though they think the subjects are important to a proper presentation of a course.

Implementation of Act 580 will have serious and untoward consequences for students, particularly those planning to attend college. Evolution is the cornerstone of modern biology, and many courses in public schools contain subject matter relating to such varied topics as the age of the earth, geology and relationships among living things. Any student who is deprived of instruction as to the prevailing scientific thought on these topics will be denied a significant part of science education. Such a deprivation through the high school level would undoubtedly have an impact upon the quality of education in the State's colleges and universities, especially including the pre-professional and professional programs in the health sciences.

"Evolution is the cornerstone of modern biology," certainly gives enormous weight to a theoretical teaching that is not supposed to be presented as dogma. While it's true that eliminating evolution from the curriculum would severely limit what can be taught, this bill did not ask for that to be done. It asked for balance, for equal time, to present the opposing viewpoint. This transcript direly predicts that instead teachers will live n fear of reprisals from wild-eyed religionists and children's education will suffer. They will be denied a full preparation and have stunted opportunities and a blighted future. Who would not fear the possibility that their children might fail to learn important things, things they need to go on in their education?

The defendants argue in their brief that evolution is, in effect, a religion, and that by teaching a religion which is contrary to some students' religious views, the State is infringing upon the student's free exercise rights under the First Amendment. Mr. Ellwanger's legislative findings, which were adopted as a finding of fact by the Arkansas Legislature in Act 590, provides:

Evolution-science is contrary to the religious convictions or moral values or philosophical beliefs of many students and parents, including individuals of many different religious faiths and with diverse moral and philosophical beliefs. Act 590, &7(d).

The defendants argue that the teaching of evolution alone presents both a free exercise problem and an establishment problem which can only be redressed by giving balanced treatment to creation science, which is admittedly consistent with some religious beliefs. This argument appears to have its genesis in a student note written by Mr. Wendell Bird, “Freedom of Religion and Science Instruction in Public Schools,” 87 Yale L.J. 515 (1978). The argument has no legal merit.

If creation science is, in fact, science and not religion, as the defendants claim, it is difficult to see how the teaching of such a science could “neutralize” the religious nature of evolution.

Assuming for the purposes of argument, however, that evolution is a religion or religious tenet, the remedy is to stop the teaching of evolution, not establish another religion in opposition to it. Yet it is clearly established in the case law, and perhaps also in common sense, that evolution is not a religion and that teaching evolution does not violate the Establishment Clause, Epperson v. Arkansas, supra, Willoughby v. Stever, No. 15574-75 (D.D.C. May 18, 1973); aff'd. 504 F.2d 271 (D.C. Cir. 1974), cert. denied , 420 U.S. 924 (1975); Wright v. Houston Indep. School Dist., 366 F. Supp. 1208 (S.D. Tex 1978), aff.d. 486 F.2d 137 (5th Cir. 1973), cert. denied 417 U.S. 969 (1974).

As proof that Evolution is not a religion, the transcript cites precedent, case law. This is circular reasoning if there ever was such a thing. The court rules that Evolution is not a religion because another court has already ruled that Evolution is not a religion. Case closed.

The Court closes this opinion with a thought expressed eloquently by the great Justice Frankfurter:

We renew our conviction that "we have at stake the very existence of our country on the faith that complete separation between the state and religion is best for the state and best for religion." Everson v. Board of Education, 330 U.S. at 59. If nowhere else, in the relation between Church and State, "good fences make good neighbors." [McCollum v. Board of Education, 333 U.S. 203, 232 (1948)]

Once again Justice Frankfurter claims for the state the sole right to educate, and to dictate what shall and shall not be allowed in the scope of that education.

The quote from Robert Frost's 1914 poem "Mending Wall" is extremely significant in this context. The narrator of the poem objects to the metaphorical walls dividing people from one another. If the state is that neighbor on the other side, I feel the same as the narrator of Frost's poem:

Before I built a wall I'd ask to know
What I was walling in or walling out,
And to whom I was like to give offense.
Something there is that doesn't love a wall,
That wants it down.'
I could say 'Elves' to him,
But it's not elves exactly, and I'd rather
He said it for himself. I see him there
Bringing a stone grasped firmly by the top
In each hand, like an old-stone savage armed.
He moves in darkness as it seems to me,
Not of woods only and the shade of trees.
He will not go behind his father's saying,
And he likes having thought of it so well
He says again,
'Good fences make good neighbors.'

Edwards v. Aguillard, U.S. Supreme Court, 1987

(Author's Note: Italicized material represents direct quotations.

Material in regular type represents the author's comments.)

(The following paragraph is not part of a trial transcript, but is quoted from the website Voices For Evolution)

In 1987, in Edwards v. Aguillard, the U.S. Supreme Court held unconstitutional Louisiana's "Creationism Act." This statute prohibited the teaching of evolution in public schools, except when it was accompanied by instruction in "creation science." The Court found that, by advancing the religious belief that a supernatural being created humankind, which is embraced by the term creation science, the act impermissibly endorses religion. In addition, the Court found that the provision of a comprehensive science education is undermined when it is forbidden to teach evolution except when creation science is also taught. (Segraves v. State of California (1981) Sacramento Superior Court #278978)

Instead of excerpts from the full trial transcript, which is essentially the same issue as that treated in the case of McLean v. Arkansas Board of Education, 1982, what follows is Justice Antonin Scalia's dissenting opinion, which was shared by the Chief Justice William Rehnquist. This material is taken from www.talkorigins.org/faqs/ edwards-v-aguillard. It is lengthy and includes many notes and references but his statement is in vocabulary easy to understand and well worth reading. It is possible to get through the notes and references with patient effort. There is no inserted commentary because the justice's statements are those of a straightforward, honest man honestly frustrated by his inability to stop an injustice. They need no real explanation and would be difficult to improve upon.

Even if I agreed with the questionable premise that legislation can be invalidated under the Establishment Clause on the basis of its motivation alone, without regard to its effects, I would still find no justification for today's decision. The Louisiana legislators who passed the "Balanced Treatment for Creation-Science and Evolution-Science Act" (Balanced Treatment Act), La. Rev. Stat. Ann. 17:286.1-17:286.7 (West 1982), each of whom had sworn to support the Constitution were well aware of the potential Establishment Clause problems and considered that aspect of the legislation with great care. After seven hearings and several months of study, resulting in substantial revision of the original proposal, they approved the Act overwhelmingly and specifically articulated the secular purpose they meant it to serve.

Although the record contains abundant evidence of the sincerity of that purpose (the only issue pertinent to this case), the Court today holds, essentially on the basis of "its visceral knowledge regarding what must have motivated the legislators," 778 F.2d 225, 227 (CA5 1985) (Gee, J., dissenting) (emphasis added), that the members of the Louisiana Legislature knowingly violated their oaths and then lied about it. I dissent. Had requirements of the Balanced Treatment Act that are not apparent on its face been clarified by an interpretation of the Louisiana Supreme Court, or by the manner of its implementation, the Act might well be found unconstitutional; but the question of its constitutionality cannot rightly be disposed of on the gallop, by impugning the motives of its supporters.

I

This case arrives here in the following posture: The Louisiana Supreme Court has never been given an opportunity to interpret the Balanced Treatment Act, State officials have never attempted to implement it, and it has never been the subject of a full evidentiary hearing. We can only guess at its meaning. We know that it forbids instruction in either "creation-science" or "evolution-science" without instruction in the other, @ 17:286.4A, but the parties are sharply divided over what creation science consists of. Appellants insist that it is a collection of educationally valuable scientific data that has been censored from classrooms by an embarrassed scientific establishment. Appellees insist it is not science at all but thinly veiled religious doctrine. Both interpretations of the intended meaning of that phrase find considerable support in the legislative history.

At least at this stage in the litigation, it is plain to me that we must accept appellants' view of what the statute means. To begin with, the statute itself defines "creation-science" as "the scientific evidences for creation and inferences from those scientific evidences." @ 17:286.3(2) (emphasis added). If, however, that definition is not thought sufficiently helpful, the means by which the Louisiana Supreme Court will give the term more precise content is quite clear -- and again, at this stage in the litigation, favors the appellants' view. "Creation science" is unquestionably a "term of art," see Brief for 72 Nobel Laureates et al. as Amici Curiae 20, and thus, under Louisiana law, is "to be interpreted according to [its] received meaning and acceptation with the learned in the art, trade or profession to which [it] refer[s]." La. Civ. Code Ann., Art. 15 (West 1952). The only evidence in the record of the "received meaning and acceptation" of "creation science" is found in five affidavits filed by appellants. In those affidavits, two scientists, a philosopher, a theologian, and an educator, all of whom claim extensive knowledge of creation science, swear that it is essentially a collection of scientific data supporting the theory that the physical universe and life within it appeared suddenly and have not changed substantially since appearing. See App. to Juris. Statement A-19 (Kenyon); id., at A-36 (Morrow); id., at A-41 (Miethe). These experts insist that creation science is a strictly scientific concept that can be presented without religious reference. See id., at A-19 -- A-20, A-35 (Kenyon); id., at A-36 -- A-38 (Morrow); id., at A-40, A-41, A-43 (Miethe); id., at A-47, A-48 (Most); id., at A-49 (Clinkert). At this point, then, we must assume that the Balanced Treatment Act does not require the presentation of religious doctrine.

Nothing in today's opinion is plainly to the contrary, but what the statute means and what it requires are of rather little concern to the Court. Like the Court of Appeals, 765 F.2d 1251, 1253, 1254 (CA5 1985), the Court finds it necessary to consider only the motives of the legislators who supported the

Balanced Treatment Act, ante, at 586, 593-594, 596. After examining the statute, its legislative history, and its historical and social context, the Court holds that the Louisiana Legislature acted without "a secular legislative purpose" and that the Act therefore fails the "purpose" prong of the three-part test set forth in Lemon v. Kurtzman, 403 U.S. 602, 612 (1971). As I explain below, infra, at 636-640, I doubt whether that "purpose" requirement of Lemon is a proper interpretation of the Constitution; but even if it were, I could not agree with the Court's assessment that the requirement was not satisfied here.

This Court has said little about the first component of the Lemon test. Almost invariably, we have effortlessly discovered a secular purpose for measures challenged under the Establishment Clause, typically devoting no more than a sentence or two to the matter. See, e. g., Witters v. Washington Dept. of Services for Blind, 474 U.S. 481, 485-486 (1986); Grand Rapids School District v. Ball, 473 U.S. 373, 383 (1985); Mueller v. Allen, 463 U.S. 388, 394-395 (1983); Larkin v. Grendel's Den, Inc., 459 U.S. 116, 123-124 (1982); Widmar v. Vincent, 454 U.S. 263, 271 (1981); Committee for Public Education & Religious Liberty v. Regan, 444 U.S. 646, 654, 657 (1980); Wolman v. Walter, 433 U.S. 229, 236 (1977) (plurality opinion); Meek v. Pittenger, 421 U.S. 349, 363 (1975); Committee for Public Education & Religious Liberty v. Nyquist, 413 U.S. 756, 773 (1973); Levitt v. Committee for Public Education & Religious Liberty, 413 U.S. 472, 479-480, n. 7 (1973); Tilton v. Richardson, 403 U.S. 672, 678-679 (1971) (plurality opinion); Lemon v. Kurtzman, supra, at 613. In fact, only once before deciding Lemon, and twice since, have we invalidated a law for lack of a secular purpose. See Wallace v. Jaffree, 472 U.S. 38 (1985); Stone v. Graham, 449 U.S. 39 (1980) (per curiam); Epperson v. Arkansas, 393 U.S. 97 (1968).

Nevertheless, a few principles have emerged from our cases, principles which should, but to an unfortunately large extent do not, guide the Court's application of Lemon today. It is clear, first of all, that regardless of what "legislative purpose" may mean in other contexts, for the purpose of the Lemon test it means the "actual" motives of those responsible for the challenged action. The Court recognizes this, see ante, at 585, as it has in the past, see, e. g., Witters v. Washington Dept. of Services for Blind, supra, at 486; Wallace v. Jaffree, supra, at 56. Thus, if those legislators who supported the Balanced Treatment Act in fact acted with a "sincere" secular purpose, ante, at 587, the Act survives the first component of the Lemon test, regardless of whether that purpose is likely to be achieved by the provisions they enacted.

Our cases have also confirmed that when the Lemon Court referred to "a secular . . . purpose," 403 U.S., at 612, it meant "a secular purpose." The author of Lemon, writing for the Court, has said that invalidation under the purpose prong is appropriate when "there [is] no question that the statute or activity was motivated wholly by religious considerations." Lynch v. Donnelly, 465 U.S. 668, 680 (1984) (Burger, C. J.) (emphasis added); see also id., at 681, n. 6; Wallace v. Jaffree, supra, at 56 ("The First Amendment requires that a statute must be invalidated if it is entirely motivated by a purpose to advance religion") (emphasis added; footnote omitted). In all three cases in which we struck down laws under the Establishment Clause for lack of a secular purpose, we found that the legislature's sole motive was to promote religion. See Wallace v. Jaffree, supra, at 56, 57, 60; Stone v. Graham, supra, at 41, 43, n. 5; Epperson v. Arkansas, supra, at 103, 107-108; see also Lynch v. Donnelly, supra, at 680 (describing Stone and Epperson as cases in which we invalidated laws "motivated wholly by religious considerations"). Thus, the majority's invalidation of the Balanced Treatment Act is defensible only if the record indicates that the Louisiana Legislature had no secular purpose.

It is important to stress that the purpose forbidden by Lemon is the purpose to "advance religion." 403 U.S., at 613; accord, ante, at 585 ("promote" religion); Witters v. Washington Dept. of Services for Blind, supra, at 486 ("endorse religion"); Wallace v. Jaffree, 472 U.S., at 56 ("advance religion"); ibid. ("endorse . . . religion"); Committee for Public Education & Religious Liberty v. Nyquist, supra, at 788 ("'advancing' . . . religion"); Levitt v. Committee for Public Education & Religious Liberty, supra, at 481 ("advancing religion"); Walz v. Tax Comm'n of New York City, 397 U.S. 664, 674 (1970) ("establishing, sponsoring, or supporting religion"); Board of Education v. Allen, 392 U.S. 236, 243 (1968) ("'advancement or inhibition of religion'") (quoting Abington School Dist. v. Schempp, 374 U.S. 203, 222 (1963)). Our cases in no way imply that the Establishment Clause forbids legislators merely to act upon their religious convictions. We surely would not strike down a law providing money to feed the hungry or shelter the homeless if it could be demonstrated that, but for the religious beliefs of the legislators, the funds would not have been approved. Also, political activism by the religiously motivated is part of our heritage. Notwithstanding the majority's implication to the contrary, ante, at 589-591, we do not presume that the sole purpose of a law is to advance religion merely because it was supported strongly by organized religions or by adherents of particular faiths. See Walz v. Tax Comm'n of New York City, supra, at 670; cf. Harris v. McRae, 448 U.S. 297, 319-320 (1980). To do so would deprive religious men and women of their right to participate in the political process. Today's religious activism may give us the Balanced Treatment Act, but yesterday's resulted in the abolition of slavery, and tomorrow's may bring relief for famine victims. Similarly, we will not presume that a law's purpose is to advance religion merely because it "'happens to coincide or harmonize with the tenets of some or all religions,'" Harris v. McRae, supra, at 319 (quoting McGowan v. Maryland, 366 U.S. 420, 442 (1961)), or because it benefits religion, even substantially. We have, for example, turned back Establishment Clause challenges to restrictions on abortion funding, Harris v. McRae, supra, and to Sunday closing laws, McGowan v. Maryland, supra, despite the fact that both "agre[e] with the dictates of [some] Judaeo-Christian religions," id., at 442. "In many instances, the Congress or state legislatures conclude that the general welfare of society, wholly apart from any religious considerations, demands such regulation." Ibid. On many past occasions we have had no difficulty finding a secular purpose for governmental action far more likely to advance religion than the Balanced Treatment Act. See, e. g., Mueller v. Allen, 463 U.S., at 394-395 (tax deduction for expenses of religious education); Wolman v. Walter, 433 U.S., at 236 (plurality opinion) (aid to religious schools); Meek v. Pittenger, 421 U.S., at 363 (same); Committee for Public Education & Religious Liberty v. Nyquist, 413 U.S., at 773 (same); Lemon v. Kurtzman, 403 U.S., at 613 (same); Walz v. Tax Comm'n of New York City, supra, at 672 (tax exemption for church property); Board of Education v. Allen, supra, at 243 (textbook loans to students in religious schools). Thus, the fact that creation science coincides with the beliefs of certain religions, a fact upon which the majority relies heavily, does not itself justify invalidation of the Act.

Finally, our cases indicate that even certain kinds of governmental actions undertaken with the specific intention of improving the position of religion do not "advance religion" as that term is used in Lemon. 403 U.S., at 613. Rather, we have said that in at least two circumstances government must act to advance religion, and that in a third it may do so.

First, since we have consistently described the Establishment Clause as forbidding not only state action motivated by the desire to advance religion, but also that intended to "disapprove," "inhibit," or evince "hostility" toward religion, see, e. g., ante, at 585 ("'disapprove'") (quoting Lynch v. Donnelly, supra, at 690 (O'CONNOR, J., concurring)); Lynch v. Donnelly, supra, at 673 ("hostility"); Committee for Public Education & Religious Liberty v. Nyquist, supra, at 788 ("'inhibi[t]'"); and since we have said that governmental “neutrality” toward religion is the preeminent goal of the First Amendment, see, e. g., Grand Rapids School District v. Ball, 473 U.S., at 382; Roemer v. Maryland Public Works Bd., 426 U.S. 736, 747 (1976) (plurality opinion); Committee for Public Education & Religious Liberty v. Nyquist, supra, at 792-793; a State which discovers that its employees are inhibiting religion must take steps to prevent them from doing so, even though its purpose would clearly be to advance religion. Cf. Walz v. Tax Comm’n of New York City, supra, at 673. Thus, if the Louisiana Legislature sincerely believed that the State’s science teachers were being hostile to religion, our cases indicate that it could act to eliminate that hostility without running afoul of Lemon’s purpose test.

Second, we have held that intentional governmental advancement of religion is sometimes required by the Free Exercise Clause. For example, in Hobbie v. Unemployment Appeals Comm’n of Fla., 480 U.S. 136 (1987); Thomas v. Review Bd., Indiana Employment Security Div., 450 U.S. 707 (1981); Wisconsin v. Yoder, 406 U.S. 205 (1972); and Sherbert v. Verner, 374 U.S. 398 (1963), we held that in some circumstances States must accommodate the beliefs of religious citizens by exempting them from generally applicable regulations. We have not yet come close to reconciling Lemon and our Free Exercise cases, and typically we do not really try. See, e. g., Hobbie v. Unemployment Appeals Comm’n of Fla., supra, at 144-145; Thomas v. Review Bd., Indiana Employment Security Div., supra, at 719-720. It is clear, however, that members of the Louisiana Legislature were not impermissibly motivated for purposes of the Lemon test if they believed that approval of the Balanced Treatment Act was required by the Free Exercise Clause.

We have also held that in some circumstances government may act to accommodate religion, even if that action is not required by the First Amendment. See Hobbie v. Unemployment Appeals Comm'n of Fla., supra, at 144-145. It is well established that "the limits of permissible state accommodation to religion are by no means co-extensive with the noninterference mandated by the Free Exercise Clause." Walz v. Tax Comm'n of New York City, supra, at 673; see also Gillette v. United States, 401 U.S. 437, 453 (1971). We have implied that voluntary governmental accommodation of religion is not only permissible, but desirable. See, e. g., ibid. Thus, few would contend that Title VII of the Civil Rights Act of 1964, which both forbids religious discrimination by private-sector employers, 78 Stat. 255, 42 U. S. C. @ 2000e-2(a)(1), and requires them reasonably to accommodate the religious practices of their employees, @ 2000e(j), violates the Establishment Clause, even though its "purpose" is, of course, to advance religion, and even though it is almost certainly not required by the Free Exercise Clause. While we have warned that at some point, accommodation may devolve into "an unlawful fostering of religion," Hobbie v. Unemployment Appeals Comm'n of Fla., supra, at 145, we have not suggested precisely (or even roughly) where that point might be. It is possible, then, that even if the sole motive of those voting for the Balanced Treatment Act was to advance religion, and its passage was not actually required, or even believed to be required, by either the Free Exercise or Establishment Clauses, the Act would nonetheless survive scrutiny under Lemon's purpose test.

One final observation about the application of that test: Although the Court's opinion gives no hint of it, in the past we have repeatedly affirmed "our reluctance to attribute unconstitutional motives to the States." Mueller v. Allen, supra, at 394; see also Lynch v. Donnelly, 465 U.S., at 699 (BRENNAN, J., dissenting). We "presume that legislatures act in a constitutional manner." Illinois v. Krull, 480 U.S. 340, 351 (1987); see also Clements v. Fashing, 457 U.S. 957, 963 (1982) (plurality opinion); Rostker v. Goldberg, 453 U.S. 57, 64

(1981); McDonald v. Board of Election Comm'rs of Chicago, 394 U.S. 802, 809 (1969). Whenever we are called upon to judge the constitutionality of an act of a state legislature, "we must have 'due regard to the fact that this Court is not exercising a primary judgment but is sitting in judgment upon those who also have taken the oath to observe the Constitution and who have the responsibility for carrying on government.'" Rostker v. Goldberg, supra, at 64 (quoting Joint Anti-Fascist Refugee Committee v. McGrath, 341 U.S. 123, 164 (1951) (Frankfurter, J., concurring)). This is particularly true, we have said, where the legislature has specifically considered the question of a law's constitutionality. Ibid.

With the foregoing in mind, I now turn to the purposes underlying adoption of the Balanced Treatment Act.

II

II A

We have relatively little information upon which to judge the motives of those who supported the Act. About the only direct evidence is the statute itself and transcripts of the seven committee hearings at which it was considered. Unfortunately, several of those hearings were sparsely attended, and the legislators who were present revealed little about their motives. We have no committee reports, no floor debates, no remarks inserted into the legislative history, no statement from the Governor, and no postenactment statements or testimony from the bill's sponsor or any other legislators. Cf. Wallace v. Jaffree, 472 U.S., at 43, 56-57. Nevertheless, there is ample evidence that the majority is wrong in holding that the Balanced Treatment Act is without secular purpose.

At the outset, it is important to note that the Balanced Treatment Act did not fly through the Louisiana Legislature on wings of fundamentalist religious fervor -- which would be unlikely, in any event, since only a small minority of the State's citizens belong to fundamentalist religious denominations. See B. Quinn, H. Anderson, M. Bradley, P. Goetting, & P. Shriver, Churches and Church Membership in the United States 16 (1982). The Act had its genesis (so to speak) in legislation introduced by Senator Bill Keith in June 1980. After two hearings before the Senate Committee on Education, Senator Keith asked that his bill be referred to a study commission composed of members of both Houses of the Louisiana Legislature. He expressed hope that the joint committee would give the bill careful consideration and determine whether his arguments were "legitimate." 1 App. E-29

-- E-30. The committee met twice during the interim, heard testimony (both for and against the bill) from several witnesses, and received staff reports. Senator Keith introduced his bill again when the legislature reconvened. The Senate Committee on Education held two more hearings and approved the bill after substantially amending it (in part over Senator Keith's objection). After approval by the full Senate, the bill was referred to the House Committee on Education. That committee conducted a lengthy hearing, adopted further amendments, and sent the bill on to the full House, where it received favorable consideration. The Senate concurred in the House amendments and on July 20, 1981, the Governor signed the bill into law.

Senator Keith's statements before the various committees that considered the bill hardly reflect the confidence of a man preaching to the converted. He asked his colleagues to "keep an open mind" and not to be "biased" by misleading characterizations of creation science. Id., at E-33. He also urged them to "look at this subject on its merits and not on some preconceived idea." Id., at E-34; see also 2 id., at E-491. Senator Keith's reception was not especially warm. Over his strenuous objection, the Senate Committee on Education voted 5-1 to amend his bill to deprive it of any force; as amended, the bill merely gave teachers permission to balance the teaching of creation science or evolution with the other. 1 id., at E-442 -- E-461. The House Committee restored the "mandatory" language to the bill by a vote of only 6-5, 2 id., at E-626 -- E-627, and both the full House (by vote of 52-35), id., at E-700 -- E-706, and full Senate (23-15), id., at E-

735 -- E-738, had to repel further efforts to gut the bill.

The legislators understood that Senator Keith's bill involved a "unique" subject, 1 id., at E-106 (Rep. M. Thompson), and they were repeatedly made aware of its potential constitutional problems, see, e. g., id., at E-26 -- E-28 (McGehee); id., at E-38 -- E-39 (Sen. Keith); id., at E-241 -- E-242 (Rossman); id., at E-257 (Probst); id., at E-261 (Beck); id., at E-282 (Sen. Keith). Although the Establishment Clause, including its secular purpose requirement, was of substantial concern to the legislators, they eventually voted overwhelmingly in favor of the Balanced Treatment Act: The House approved it 71-19 (with 15 members absent), 2 id., at E-716 -- E-722; the Senate 26-12 (with all members present), id., at E-741 -- E-744. The legislators specifically designated the protection of "academic freedom" as the purpose of the Act. La. Rev. Stat. Ann. @ 17:286.2 (West 1982). We cannot accurately assess whether this purpose is a "sham," ante, at 587, until we first examine the evidence presented to the legislature far more carefully than the Court has done.

Before summarizing the testimony of Senator Keith and his supporters, I wish to make clear that I by no means intend to endorse its accuracy. But my views (and the views of this Court) about creation science and evolution are (or should be) beside the point. Our task is not to judge the debate about teaching the origins of life, but to ascertain what the members of the Louisiana Legislature believed. The vast majority of them voted to approve a bill which explicitly stated a secular purpose; what is crucial is not their wisdom in believing that purpose would be achieved by the bill, but their sincerity in believing it would be.

Most of the testimony in support of Senator Keith's bill came from the Senator himself and from scientists and educators he presented, many of whom enjoyed academic credentials that may have been regarded as quite impressive by members of the Louisiana Legislature. To a substantial extent, their testimony was devoted to lengthy, and, to the layman, seemingly expert scientific expositions on the origin of life. See, e. g., 1 App. E-11 -- E-18 (Sunderland); id., at E-50 -- E-60 (Boudreaux); id., at E-86 -- E-89 (Ward); id., at E-130 -- E-153 (Boudreaux paper); id., at E-321 -- E-326 (Boudreaux); id., at E-423 -- E-428 (Sen. Keith). These scientific lectures touched upon, inter alia, biology, paleontology, genetics, astronomy, astrophysics, probability analysis, and biochemistry. The witnesses repeatedly assured committee members that "hundreds and hundreds" of highly respected, internationally renowned scientists believed in creation science and would support their testimony. See, e. g., id., at E-5 (Sunderland); id., at E-76 (Sen. Keith); id., at E-100 -- E-101 (Reiboldt); id., at E-327 -- E-328 (Boudreaux); 2 id., at E-503 -- E-504 (Boudreaux).

Senator Keith and his witnesses testified essentially as set forth in the following numbered paragraphs:

(1) There are two and only two scientific explanations for the beginning of life -- evolution and creation science. 1 id., at E-6 (Sunderland); id., at E-34 (Sen. Keith); id., at E-280 (Sen. Keith); id., at E-417 -- E-418 (Sen. Keith). Both are bona fide

"sciences." Id., at E-6 -- E-7 (Sunderland); id., at E-12 (Sunderland); id., at E-416 (Sen. Keith); id., at E-427 (Sen. Keith); 2 id., at E-491 -- E-492 (Sen. Keith); id., at E-497 -- E-498 (Sen. Keith). Both posit a theory of the origin of life and subject that theory to empirical testing. Evolution posits that life arose out of inanimate chemical compounds and has gradually evolved over millions of years. Creation science posits that all life forms now on earth appeared suddenly and relatively recently and have changed little. Since there are only two possible explanations of the origin of life, any evidence that tends to disprove the theory of evolution necessarily tends to prove the theory of creation science, and vice versa. For example, the abrupt appearance in the fossil record of complex life, and the extreme rarity of transitional life forms in that record, are evidence for creation science. 1 id., at E-7 (Sunderland); id., at E-12 -- E-18 (Sunderland); id., at E-45 -- E-60 (Boudreaux); id., at E-67 (Harlow); id., at E-130 -- E-153 (Boudreaux paper); id., at E-423 -- E-428 (Sen. Keith).

(2) The body of scientific evidence supporting creation science is as strong as that supporting evolution. In fact, it may be stronger. Id., at E-214 (Young statement); id., at E-310 (Sen. Keith); id., at E-416 (Sen. Keith); 2 id., at E-492 (Sen. Keith). The evidence for evolution is far less compelling than we have been led to believe. Evolution is not a scientific "fact," since it cannot actually be observed in a laboratory. Rather, evolution is merely a scientific theory or "guess." 1 id., at E-20 -- E-21 (Morris); id., at E-85 (Ward); id., at E-100 (Reiboldt); id., at E-328 -- E-329 (Boudreaux); 2 id., at E-506 (Boudreaux). It is a very bad guess at that. The

scientific problems with evolution are so serious that it could accurately be termed a "myth." 1 id., at E-85 (Ward); id., at E-92 -- E-93 (Kalivoda); id., at E-95 -- E-97 (Sen. Keith); id., at E-154 (Boudreaux paper); id., at E-329 (Boudreaux); id., at E-453 (Sen. Keith); 2 id., at E-505 -- E-506 (Boudreaux); id., at E-516 (Young).

(3) Creation science is educationally valuable. Students exposed to it better understand the current state of scientific evidence about the origin of life. 1 id., at E-19 (Sunderland); id., at E-39 (Sen. Keith); id., at E-79 (Kalivoda); id., at E-308 (Sen. Keith); 2 id., at E-513 -- E-514 (Morris). Those students even have a better understanding of evolution. 1 id., at E-19 (Sunderland). Creation science can and should be presented to children without any religious content. Id., at E-12 (Sunderland); id., at E-22 (Sanderford); id., at E-35 -- E-36 (Sen. Keith); id., at E-101 (Reiboldt); id., at E-279 -- E-280 (Sen. Keith); id., at E-282 (Sen. Keith).

(4) Although creation science is educationally valuable and strictly scientific, it is now being censored from or misrepresented in the public schools. Id., at E-19 (Sunderland); id., at E-21 (Morris); id., at E-34 (Sen. Keith); id., at E-37 (Sen. Keith); id., at E-42 (Sen. Keith); id., at E-92 (Kalivoda); id., at E-97 -- E-98 (Reiboldt); id., at E-214 (Young statement); id., at E-218 (Young statement); id., at E-280 (Sen. Keith); id., at E-309 (Sen. Keith); 2 id., at E-513 (Morris). Evolution, in turn, is misrepresented as an absolute truth. 1 id., at E-63 (Harlow); id., at E-74 (Sen. Keith); id., at E-81 (Kalivoda); id., at E-214 (Young statement); 2 id., at E-507 (Harlow); id., at E-513 (Morris); id., at E-516

(Young). Teachers have been brainwashed by an entrenched scientific establishment composed almost exclusively of scientists to whom evolution is like a “religion.” These scientists discriminate against creation scientists so as to prevent evolution’s weaknesses from being exposed. 1 id., at E-61 (Boudreaux); id., at E-63 -- E-64 (Harlow); id., at E-78 -- E-79 (Kalivoda); id., at E-80 (Kalivoda); id., at E-95 -- E-97 (Sen. Keith); id., at E-129 (Boudreaux paper); id., at E-218 (Young statement); id., at E-357 (Sen. Keith); id., at E-430 (Boudreaux).

(5) The censorship of creation science has at least two harmful effects. First, it deprives students of knowledge of one of the two scientific explanations for the origin of life and leads them to believe that evolution is proven fact; thus, their education suffers and they are wrongly taught that science has proved their religious beliefs false. Second, it violates the Establishment Clause. The United States Supreme Court has held that secular humanism is a religion. Id., at E-36 (Sen. Keith) (referring to Torcaso v. Watkins, 367 U.S. 488, 495, n. 11 (1961)); 1 App. E-418 (Sen. Keith); 2 id., at E-499 (Sen. Keith). Belief in evolution is a central tenet of that religion. 1 id., at E-282 (Sen. Keith); id., at E-312 -- E-313 (Sen. Keith); id., at E-317 (Sen. Keith); id., at E-418 (Sen. Keith); 2 id., at E-499 (Sen. Keith). Thus, by censoring creation science and instructing students that evolution is fact, public school teachers are now advancing religion in violation of the Establishment Clause. 1 id., at E-2 -- E-4 (Sen. Keith); id., at E-36 -- E-37, E-39 (Sen. Keith); id., at E-154 -- E-155 (Boudreaux paper); id., at E-281 -- E-282 (Sen. Keith); id., at E-313 (Sen. Keith); id., at E-315 -- E-316 (Sen. Keith); id., at E-

317 (Sen. Keith); 2 id., at E-499 -- E-500 (Sen. Keith).

Senator Keith repeatedly and vehemently denied that his purpose was to advance a particular religious doctrine. At the outset of the first hearing on the legislation, he testified: "We are not going to say today that you should have some kind of religious instructions in our schools. . . . We are not talking about religion today. . . . I am not proposing that we take the Bible in each science class and read the first chapter of Genesis." 1 id., at E-35. At a later hearing, Senator Keith stressed: "To . . . teach religion and disguise it as creationism . . . is not my intent. My intent is to see to it that our textbooks are not censored." Id., at E-280. He made many similar statements throughout the hearings. See, e. g., id., at E-41; id., at E-282; id., at E-310; id., at E-417; see also id., at E-44 (Boudreaux); id., at E-80 (Kalivoda).

We have no way of knowing, of course, how many legislators believed the testimony of Senator Keith and his witnesses. But in the absence of evidence to the contrary (4), we have to assume that many of them did. Given that assumption, the Court today plainly errs in holding that the Louisiana Legislature passed the Balanced Treatment Act for exclusively religious purposes.

II B

Even with nothing more than this legislative history to go on, I think it would be extraordinary to invalidate the Balanced Treatment Act for lack of a valid secular purpose. Striking down a law approved by the democratically elected representatives of the people is no minor matter. "The cardinal principle of statutory construction is to save and not to destroy. We have repeatedly held that as between two possible interpretations of a statute, by one of which it would be unconstitutional and by the other valid, our plain duty is to adopt that which will save the act." NLRB v. Jones & Laughlin Steel Corp., 301 U.S. 1, 30 (1937). So, too, it seems to me, with discerning statutory purpose. Even if the legislative history were silent or ambiguous about the existence of a secular purpose -- and here it is not -- the statute should survive Lemon's purpose test. But even more validation than mere legislative history is present here. The Louisiana Legislature explicitly set forth its secular purpose ("protecting academic freedom") in the very text of the Act. La. Rev. Stat. @ 17:286.2 (West 1982). We have in the past repeatedly relied upon or deferred to such expressions, see, e. g., Committee for Public Education & Religious Liberty v. Regan, 444 U.S., at 654; Meek v. Pittenger, 421 U.S., at 363, 367-368; Committee for Public Education & Religious Liberty v. Nyquist, 413 U.S., at 773; Levitt v. Committee for Public Education & Religious Liberty, 413 U.S., at 479-480, n. 7; Tilton v. Richardson, 403 U.S., at 678-679 (plurality opinion); Lemon v. Kurtzman, 403 U.S., at 613; Board of Education v. Allen, 392 U.S., at 243.

The Court seeks to evade the force of this expression of purpose by stubbornly misinterpreting it, and then finding that the provisions of the Act do not advance that misinterpreted purpose, thereby showing it to be a sham. The Court first surmises that "academic freedom" means "enhancing the freedom of teachers to teach what they will," ante, at 586 -- even though "academic freedom" in that sense has little scope in the structured elementary and secondary curriculums with which the Act is concerned. Alternatively, the Court suggests that it might mean "maximiz[ing] the comprehensiveness and effectiveness of science instruction," ante, at 588 -- though that is an exceedingly strange interpretation of the words, and one that is refuted on the very face of the statute. See @ 17:286.5. Had the Court devoted to this central question of the meaning of the legislatively expressed purpose a small fraction of the research into legislative history that produced its quotations of religiously motivated statements by individual legislators, it would have discerned quite readily what "academic freedom" meant: students' freedom from indoctrination. The legislature wanted to ensure that students would be free to decide for themselves how life began, based upon a fair and balanced presentation of the scientific evidence -- that is, to protect "the right of each [student] voluntarily to determine what to believe (and what not to believe) free of any coercive pressures from the State." Grand Rapids School District v. Ball, 473 U.S., at 385. The legislature did not care whether the topic of origins was taught; it simply wished to ensure that when the topic was taught, students would receive "'all of the evidence.'" Ante, at 586 (quoting Tr. of Oral Arg. 60).

As originally introduced, the "purpose" section of the Balanced Treatment Act read: "This Chapter is enacted for the purposes of protecting academic freedom . . . of students . . . and assisting students in their search for truth." 1 App. E-292 (emphasis added). Among the proposed findings of fact contained in the original version of the bill was the following: "Public school instruction in only evolution-science . . . violates the principle of academic freedom because it denies students a choice between scientific models and instead indoctrinates them in evolution science alone." Id., at E-295 (emphasis added). Senator Keith unquestionably understood "academic freedom" to mean "freedom from indoctrination." See id., at E-36 (purpose of bill is "to protect academic freedom by providing student choice"); id., at E-283 (purpose of bill is to protect "academic freedom" by giving students a "choice" rather than subjecting them to "indoctrination on origins")

If one adopts the obviously intended meaning of the statutory term "academic freedom," there is no basis whatever for concluding that the purpose they express is a "sham." Ante, at 587. To the contrary, the Act pursues that purpose plainly and consistently. It requires that, whenever the subject of origins is covered, evolution be "taught as a theory, rather than as proven scientific fact" and that scientific evidence inconsistent with the theory of evolution (viz., "creation science") be taught as well. La. Rev. Stat. Ann. @ 17:286.4A (West 1982). Living up to its title of "Balanced Treatment for Creation-Science and Evolution-Science Act," @ 17.286.1, it treats the teaching of creation the same way. It does not mandate instruction in creation science, @ 17:286.5; forbids teachers to present creation science "as proven scientific fact," @ 17:286.4A; and bans the teaching of creation science unless the theory is (to use the Court's terminology) "discredit[ed] '. . . at every turn'" with the teaching of evolution. Ante, at 589 (quoting 765 F.2d, at 1257). It surpasses understanding how the Court can see in this a purpose "to restructure the science curriculum to conform with a particular religious viewpoint," ante, at 593,"to provide a persuasive advantage to a particular religious doctrine," ante, at 592,"to promote the theory of creation science which embodies a particular religious tenet," ante, at 593, and "to endorse a particular religious doctrine," ante, at 594.

The Act's reference to "creation" is not convincing evidence of religious purpose. The Act defines creation science as "scientific evidenc[e]," @ 17:286.3(2) (emphasis added), and Senator Keith and his witnesses repeatedly stressed that the subject can and should be presented without religious content. See supra, at 623. We have no basis on the record to conclude that creation science need be anything other than a collection of scientific data supporting the theory that life abruptly appeared on earth. See n. 4, supra. Creation science, its proponents insist, no more must explain whence life came than evolution must explain whence came the inanimate materials from which it says life evolved. But even if that were not so, to posit a past creator is not to posit the eternal and personal God who is the object of religious veneration. Indeed, it is not even to posit the "unmoved mover" hypothesized by Aristotle and other notably nonfundamentalist philosophers. Senator Keith suggested this when he referred to "a creator however you define a creator." 1 App. E-280 (emphasis added).

The Court cites three provisions of the Act which, it argues, demonstrate a "discriminatory preference for the teaching of creation science" and no interest in "academic freedom." Ante, at 588. First, the Act prohibits discrimination only against creation scientists and those who teach creation science. @ 17:286.4C. Second, the Act requires local school boards to develop and provide to science teachers "a curriculum guide on presentation of creation-science." @ 17:286.7A. Finally, the Act requires the Governor to designate seven creation scientists who shall, upon request, assist local school boards in developing the curriculum guides. @ 17:286.7B. But none of these provisions casts doubt upon the sincerity of the legislators' articulated purpose of "academic freedom" -- unless, of course, one gives that term the obviously erroneous meanings preferred by the Court. The Louisiana legislators had been told repeatedly that creation scientists were scorned by most educators and scientists, who themselves had an almost religious faith in evolution. It is hardly surprising, then, that in seeking to achieve a balanced,"nonindoctrinating" curriculum, the legislators protected from discrimination only those teachers whom they thought were suffering from discrimination. (Also, the legislators were undoubtedly aware of Epperson v. Arkansas, 393 U.S. 97 (1968), and thus could quite reasonably have concluded that discrimination against evolutionists was already prohibited.) The two provisions respecting the development of curriculum guides are also consistent with "academic freedom" as the Louisiana Legislature understood the term. Witnesses had informed the legislators that, because of the hostility of most scientists and educators to creation science, the topic had been censored from or badly misrepresented in elementary and secondary school texts. In light of the unavailability of works on creation science suitable for classroom use (a fact appellees concede, see Brief for Appellees 27, 40) and the existence of ample materials on evolution, it was entirely reasonable for the legislature to conclude that science teachers attempting to implement the Act would need a curriculum guide on creation science, but not on evolution, and that those charged with developing the guide would need an easily accessible group of creation scientists. Thus, the provisions of the Act of so much concern to the Court support the conclusion that the legislature acted to advance "academic freedom."

The legislative history gives ample evidence of the sincerity of the Balanced Treatment Act's articulated purpose. Witness after witness urged the legislators to support the Act so that students would not be "indoctrinated" but would instead be free to decide for themselves, based upon a fair presentation of the scientific evidence, about the origin of life. See, e. g., 1 App. E-18 (Sunderland) ("all that we are advocating" is presenting "scientific data" to students and "letting [them] make up their own mind[s]"); id., at E-19 -- E-20 (Sunderland) (Students are now being "indoctrinated" in evolution through the use of "censored school books. . . . All that we are asking for is [the] open unbiased education in the classroom . . . your students deserve"); id., at E-21 (Morris) ("A student cannot [make an intelligent decision about the origin of life] unless he is well informed about both [evolution and creation science]"); id., at E-22 (Sanderford) ("We are asking very simply [that] . . . creationism [be presented] alongside . . . evolution and let people make their own mind[s] up"); id., at E-23 (Young) (the bill would require teachers to live up to their "obligation to present all theories" and thereby enable "students to make judgments themselves"); id., at E-44 (Boudreaux) ("Our intention is truth and as a scientist, I am interested in truth"); id., at E-60 -- E-61 (Boudreaux) ("We [teachers] are guilty of a lot of brainwashing. . . . We have a duty to . . . [present the] truth" to students "at all levels from gradeschool on through the college level"); id., at E-79 (Kalivoda) ("This [hearing] is being held I think to determine whether children will benefit from freedom of information or if they will be handicapped educationally by having little or no information about creation"); id., at E-80 (Kalivoda) ("I am not interested in teaching religion in schools. . . . I am interested in the truth and [students] having the opportunity to hear more than one side"); id., at E-98 (Reiboldt) ("The students have a right to know there is an alternate creationist point of view. They have a right to know the scientific evidences which suppor[t] that alternative"); id., at E-218 (Young statement) (passage of the bill will ensure that "communication of scientific ideas and discoveries may be unhindered"); 2 id., at E-514 (Morris) ("Are we going to allow [students] to look at evolution, to look at creationism, and to let one or the other stand or fall on its own merits, or will we by failing to pass this bill . . . deny students an opportunity to hear another viewpoint?"); id., at E-516 -- E-517 (Young) ("We want to give the children here in this state an equal opportunity to see both sides of the theories"). Senator Keith expressed similar views. See, e. g., 1 id., at E-36; id., at E-41; id., at E-280; id., at E-283.

Legislators other than Senator Keith made only a few statements providing insight into their motives, but those statements cast no doubt upon the sincerity of the Act's articulated purpose. The legislators were concerned primarily about the manner in which the subject of origins was presented in Louisiana schools -- specifically, about whether scientifically valuable information was being censored and students misled about evolution. Representatives Cain, Jenkins, and F. Thompson seemed impressed by the scientific evidence presented in support of creation science. See 2 id., at E-530 (Rep. F. Thompson); id., at E-533 (Rep. Cain); id., at E-613 (Rep. Jenkins). At the first study commission hearing, Senator Picard and Representative M. Thompson questioned Senator Keith about Louisiana teachers' treatment of evolution and creation science. See 1 id., at E-71 -- E-74. At the close of the hearing, Representative M. Thompson told the audience:

"We as members of the committee will also receive from the staff information of what is currently being taught in the Louisiana public schools. We really want to see [it]. I . . . have no idea in what manner [biology] is presented and in what manner the creationist theories [are] excluded in the public school[s]. We want to look at what the status of the situation is." Id., at E-104.

Legislators made other comments suggesting a concern about censorship and misrepresentation of scientific information. See, e. g., id., at E-386 (Sen. McLeod); 2 id., at E-527 (Rep. Jenkins); id., at E-528 (Rep. M. Thompson); id., at E-534 (Rep. Fair).

It is undoubtedly true that what prompted the legislature to direct its attention to the misrepresentation of evolution in the schools (rather than the inaccurate presentation of other topics) was its awareness of the tension between evolution and the religious beliefs of many children. But even appellees concede that a valid secular purpose is not rendered impermissible simply because its pursuit is prompted by concern for religious sensitivities. Tr. of Oral Arg. 43, 56. If a history teacher falsely told her students that the bones of Jesus Christ had been discovered, or a physics teacher that the Shroud of Turin had been conclusively established to be inexplicable on the basis of natural causes, I cannot believe (despite the majority's implication to the contrary, see ante, at 592-593) that legislators or school board members would be constitutionally prohibited from taking corrective action, simply because that action was prompted by concern for the religious beliefs of the misinstructed students.

In sum, even if one concedes, for the sake of argument, that a majority of the Louisiana Legislature voted for the Balanced Treatment Act partly in order to foster (rather than merely eliminate discrimination against) Christian fundamentalist beliefs, our cases establish that that alone would not suffice to invalidate the Act, so long as there was a genuine secular purpose as well. We have, moreover, no adequate basis for disbelieving the secular purpose set forth in the Act itself, or for concluding that it is a sham enacted to conceal the legislators' violation of their oaths of office. I am astonished by the Court's unprecedented readiness to reach such a conclusion, which I can only attribute to an intellectual predisposition created by the facts and the legend of Scopes v. State, 154 Tenn. 105, 289 S. W. 363 (1927) -- an instinctive reaction that any governmentally imposed requirements bearing upon the teaching of evolution must be a manifestation of Christian fundamentalist repression. In this case, however, it seems to me the Court's position is the repressive one. The people of Louisiana, including those who are Christian fundamentalists, are quite entitled, as a secular matter, to have whatever scientific evidence there may be against evolution presented in their schools, just as Mr. Scopes was entitled to present whatever scientific evidence there was for it. Perhaps what the Louisiana Legislature has done is unconstitutional because there is no such evidence, and the scheme they have established will amount to no more than a presentation of the Book of Genesis. But we cannot say that on the evidence before us in this summary judgment context, which includes ample uncontradicted testimony that "creation science" is a body of scientific knowledge rather than revealed belief. Infinitely less can we say (or should we say) that the scientific evidence for evolution is so conclusive that no one could be gullible enough to believe that there is any real scientific evidence to the contrary, so that the legislation's stated purpose must be a lie. Yet that illiberal judgment, that Scopes-in-reverse, is ultimately the basis on which the Court's facile rejection of the Louisiana Legislature's purpose must rest.

Since the existence of secular purpose is so entirely clear, and thus dispositive, I will not go on to discuss the fact that, even if the Louisiana

Legislature's purpose were exclusively to advance religion, some of the well-established exceptions to the impermissibility of that purpose might be applicable -- the validating intent to eliminate a perceived discrimination against a particular religion, to facilitate its free exercise, or to accommodate it. See supra, at 617-618. I am not in any case enamored of those amorphous exceptions, since I think them no more than unpredictable correctives to what is (as the next Part of this opinion will discuss) a fundamentally unsound rule. It is surprising, however, that the Court does not address these exceptions, since the context of the legislature's action gives some reason to believe they may be applicable. (6)

Because I believe that the Balanced Treatment Act had a secular purpose, which is all the first component of the Lemon test requires, I would reverse the judgment of the Court of Appeals and remand for further consideration.

III

I have to this point assumed the validity of the Lemon "purpose" test. In fact, however, I think the pessimistic evaluation that THE CHIEF JUSTICE made of the totality of Lemon is particularly applicable to the "purpose" prong: it is "a constitutional theory [that] has no basis in the history of the amendment it seeks to interpret, is difficult to apply and yields unprincipled results ..." Wallace v. Jaffree, 472 U.S., at 112 (REHNQUIST, J., dissenting).

Our cases interpreting and applying the purpose test have made such a maze of the Establishment Clause that even the most conscientious governmental officials can only guess what motives will be held unconstitutional. We have said essentially the following: Government may not act with the purpose of advancing religion, except when forced to do so by the Free Exercise Clause (which is now and then); or when eliminating existing governmental hostility to religion (which exists sometimes); or even when merely accommodating governmentally uninhibited religious practices, except that at some point (it is unclear where) intentional accommodation results in the fostering of religion, which is of course unconstitutional. See supra, at 614-618.

But the difficulty of knowing what vitiating purpose one is looking for is as nothing compared with the difficulty of knowing how or where to find it. For while it is possible to discern the objective "purpose" of a statute (i. e., the public good at which its provisions appear to be directed), or even the formal motivation for a statute where that is explicitly set forth (as it was, to no avail, here), discerning the subjective motivation of those enacting the statute is, to be honest, almost always an impossible task. The number of possible motivations, to begin with, is not binary, or indeed even finite. In the present case, for example, a particular legislator need not have voted for the Act either because he wanted to foster religion or because he wanted to improve education. He may have thought the bill would provide jobs for his district, or may have wanted to make amends with a faction of his party he had alienated on another vote, or he may have been a close friend of the bill's sponsor, or he may have been repaying a favor he owed the Majority Leader, or he may have hoped the Governor would appreciate his vote and make a fundraising appearance for him, or he may have been pressured to vote for a bill he disliked by a wealthy contributor or by a flood of constituent mail, or he may have been seeking favorable publicity, or he may have been reluctant to hurt the feelings of a loyal staff member who worked on the bill, or he may have been settling an old score with a legislator who opposed the bill, or he may have been mad at his wife who opposed the bill, or he may have been intoxicated and utterly unmotivated when the vote was called, or he may have accidentally voted "yes" instead of "no," or, of course, he may have had (and very likely did have) a combination of some of the above and many other motivations. To look for the sole purpose of even a single legislator is probably to look for something that does not exist.

Putting that problem aside, however, where ought we to look for the individual legislator's purpose? We cannot of course assume that every member present (if, as is unlikely, we know who or even how many they were) agreed with the motivation expressed in a particular legislator's preenactment floor or committee statement. Quite obviously, "what motivates one legislator to make a speech about a statute is not necessarily what motivates scores of others to enact it." United States v. O'Brien, 391 U.S. 367, 384 (1968). Can we assume, then, that they all agree with the motivation expressed in the staff-prepared committee reports they might have read -- even though we are unwilling to assume that they agreed with the motivation expressed in the very statute that they voted for? Should we consider postenactment floor statements? Or postenactment testimony from legislators, obtained expressly for the lawsuit? Should we consider media reports on the realities of the legislative bargaining? All of these sources, of course, are eminently manipulable. Legislative histories can be contrived and sanitized, favorable media coverage orchestrated, and postenactment recollections conveniently distorted. Perhaps most valuable of all would be more objective indications -- for example, evidence regarding the individual legislators' religious affiliations. And if that, why not evidence regarding the fervor or tepidity of their beliefs?

Having achieved, through these simple means, an assessment of what individual legislators intended, we must still confront the question (yet to be addressed in any of our cases) how many of them must have the invalidating intent. If a state senate approves a bill by vote of 26 to 25, and only one of the 26 intended solely to advance religion, is the law unconstitutional? What if 13 of the 26 had that intent? What if 3 of the 26 had the impermissible intent, but 3 of the 25 voting against the bill were motivated by religious hostility or were simply attempting to "balance" the votes of their impermissibly motivated colleagues? Or is it possible that the intent of the bill's sponsor is alone enough to invalidate it -- on a theory, perhaps, that even though everyone else's intent was pure, what they produced was the fruit of a forbidden tree?

Because there are no good answers to these questions, this Court has recognized from Chief Justice Marshall, see Fletcher v. Peck, 6 Cranch 87, 130 (1810), to Chief Justice Warren, United States v.

O'Brien, supra, at 383-384, that determining the subjective intent of legislators is a perilous enterprise. See also Palmer v. Thompson, 403 U.S. 217, 224-225 (1971); Epperson v. Arkansas, 393 U.S., at 113 (Black, J., concurring). It is perilous, I might note, not just for the judges who will very likely reach the wrong result, but also for the legislators who find that they must assess the validity of proposed legislation -- and risk the condemnation of having voted for an unconstitutional measure -- not on the basis of what the legislation contains, nor even on the basis of what they themselves intend, but on the basis of what others have in mind.

Given the many hazards involved in assessing the subjective intent of governmental decision makers, the first prong of Lemon is defensible, I think, only if the text of the Establishment Clause demands it. That is surely not the case. The Clause states that "Congress shall make no law respecting an establishment of religion." One could argue, I suppose, that any time Congress acts with the intent of advancing religion, it has enacted a "law respecting an establishment of religion"; but far from being an unavoidable reading, it is quite an unnatural one. I doubt, for example, that the Clayton Act, 38 Stat. 730, as amended, 15 U. S. C. @ 12 et seq., could reasonably be described as a "law respecting an establishment of religion" if bizarre new historical evidence revealed that it lacked a secular purpose, even though it has no discernible nonsecular effect. It is, in short, far from an inevitable reading of the Establishment Clause that it forbids all governmental action intended to advance religion; and if not inevitable, any reading with such untoward consequences must be wrong.

In the past we have attempted to justify our embarrassing Establishment Clause jurisprudence (7) on the ground that it "sacrifices clarity and predictability for flexibility. " Committee for Public Education & Religious Liberty v. Regan, 444 U.S., at 662. One commentator has aptly characterized this as "a euphemism . . . for . . . the absence of any principled rationale." Choper, supra n. 7, at 681. I think it time that we sacrifice some "flexibility" for "clarity and predictability." Abandoning Lemon's purpose test -- a test which exacerbates the tension between the Free Exercise and Establishment Clauses, has no basis in the language or history of the Amendment, and, as today's decision shows, has wonderfully flexible consequences -- would be a good place to start.

Notes:

1. Article VI, cl. 3, of the Constitution provides that "the Members of the several State Legislatures . . . shall be bound by Oath or Affirmation, to support this Constitution."

2. Thus the popular dictionary definitions cited by JUSTICE POWELL, ante, at 598-599 (concurring opinion), and appellees, see Brief for Appellees 25, 26; Tr. of Oral Arg. 32, 34, are utterly irrelevant, as are the views of the school superintendents cited by the majority, ante, at 595, n. 18. Three-quarters of those surveyed had "no" or "limited" knowledge of "creation-science theory," and not a single superintendent claimed "extensive" knowledge of the subject. 2 App. E-798.

3. Although creation scientists and evolutionists also disagree about the origin of the physical universe, both proponents and opponents of Senator Keith's bill focused on the question of the beginning of life.

4. Although appellees and amici dismiss the testimony of Senator Keith and his witnesses as pure fantasy, they did not bother to submit evidence of that to the District Court, making it difficult for us to agree with them. The State, by contrast, submitted the affidavits of two scientists, a philosopher, a theologian, and an educator, whose academic credentials are rather impressive. See App. to Juris. Statement A-17 -- A-18 (Kenyon); id., at A-36 (Morrow); id., at A-39 -- A-40 (Miethe); id., at A-46 -- A-47 (Most); id., at A-49 (Clinkert). Like Senator Keith and his witnesses, the affiants swear that evolution and creation science are the only two scientific explanations for the origin of life, see id., at A-19 -- A-20 (Kenyon); id., at A-38 (Morrow); id., at A-41 (Miethe); that creation science is strictly scientific, see id., at A-18 (Kenyon); id., at A-36 (Morrow); id., at A-40 -- A-41 (Miethe); id., at A-49 (Clinkert); that creation science is simply a collection of scientific data that supports the hypothesis that life appeared on earth suddenly and has changed little, see id., at A-19 (Kenyon); id., at A-36 (Morrow); id., at A-41 (Miethe); that hundreds of respected scientists believe in creation science, see id., at A-20 (Kenyon); that evidence for creation science is as strong as evidence for evolution, see id., at A-21 (Kenyon); id., at A-34 -- A-35 (Kenyon); id., at A-37 -- A-38 (Morrow); that creation science is educationally valuable, see id., at A-19 (Kenyon); id., at A-36 (Morrow); id., at A-38 -- A-39

(Morrow); id., at A-49 (Clinkert); that creation science can be presented without religious content, see id., at A-19 (Kenyon); id., at A-35 (Kenyon); id., at A-36 (Morrow); id., at A-40 (Miethe); id., at A-43 -- A-44 (Miethe); id., at A-47 (Most); id., at A-49 (Clinkert); and that creation science is now censored from classrooms while evolution is misrepresented as proven fact, see id., at A-20 (Kenyon); id., at A-35 (Kenyon); id., at A-39 (Morrow); id., at A-50 (Clinkert). It is difficult to conclude on the basis of these affidavits -- the only substantive evidence in the record -- that the laymen serving in the Louisiana Legislature must have disbelieved Senator Keith or his witnesses.

5. The majority finds it “astonishing” that I would cite a portion of Senator Keith’s original bill that was later deleted as evidence of the legislature’s understanding of the phrase “academic freedom.” Ante, at 589, n. 8. What is astonishing is the majority’s implication that the deletion of that section deprives it of value as a clear indication of what the phrase meant -- there and in the other, retained, sections of the bill. The Senate Committee on Education deleted most of the lengthy “purpose” section of the bill (with Senator Keith’s consent) because it resembled legislative “findings of fact,” which, committee members felt, should generally not be incorporated in legislation. The deletion had absolutely nothing to do with the manner in which the section described “academic freedom.” See 1 App. E-314 -- E-320; id., at E-440 -- E-442.

6. As the majority recognizes, ante, at 592, Senator Keith sincerely believed that “secular humanism is a bona fide religion,” 1 App. E-36; see also id., at E-

418; 2 id., at E-499, and that "evolution is the cornerstone of that religion," 1 id., at E-418; see also id., at E-282; id., at E-312 -- E-313; id., at E-317; 2 id., at E-499. The Senator even told his colleagues that this Court had "held" that secular humanism was a religion. See 1 id., at E-36, id., at E-418; 2 id., at E-499. (In Torcaso v. Watkins, 367 U.S. 488, 495, n. 11 (1961), we did indeed refer to "Secular Humanism" as a "religio[n].") Senator Keith and his supporters raised the "religion" of secular humanism not, as the majority suggests, to explain the source of their "disdain for the theory of evolution," ante, at 592, but to convince the legislature that the State of Louisiana was violating the Establishment Clause because its teachers were misrepresenting evolution as fact and depriving students of the information necessary to question that theory. 1 App. E-2 -- E-4 (Sen. Keith); id., at E-36 -- E-37, E-39 (Sen. Keith); id., at E-154 -- E-155 (Boudreaux paper); id., at E-281 -- E-282 (Sen. Keith); id., at E-317 (Sen. Keith); 2 id., at E-499 -- E-500 (Sen. Keith). The Senator repeatedly urged his colleagues to pass his bill to remedy this Establishment Clause violation by ensuring state neutrality in religious matters, see, e. g., 1 id., at E-36; id., at E-39; id., at E-313, surely a permissible purpose under Lemon. Senator Keith's argument may be questionable, but nothing in the statute or its legislative history gives us reason to doubt his sincerity or that of his supporters.

7. Professor Choper summarized our school aid cases thusly:

"[A] provision for therapeutic and diagnostic health services to parochial school pupils by public employees is invalid if provided in the parochial school, but not if offered at a neutral site, even if in a mobile unit adjacent to the parochial school. Reimbursement to parochial schools for the expense of administering teacher-prepared tests required by state law is invalid, but the state may reimburse parochial schools for the expense of administering state-prepared tests. The state may lend school textbooks to parochial school pupils because, the Court has explained, the books can be checked in advance for religious content and are 'self-policing'; but the state may not lend other seemingly self-policing instructional items such as tape recorders and maps. The state may pay the cost of bus transportation to parochial schools, which the Court has ruled are 'permeated' with religion; but the state is forbidden to pay for field trip transportation visits 'to governmental, industrial, cultural, and scientific centers designed to enrich the secular studies of students.'" Choper, The Religion Clauses of the First Amendment: Reconciling the Conflict, 41 U. Pitt. L. Rev. 673, 680-681 (1980) (footnotes omitted).

Since that was written, more decisions on the subject have been rendered, but they leave the theme of chaos securely unimpaired. See, e. g., Aguilar v.Felton, 473 U.S. 402 (1985); Grand Rapids School District v. Ball, 473 U.S. 373 (1985).

Webster v. New Lenox School District, 1990, the Seventh Circuit Court of Appeals

(Author's Note: Italicized material represents direct quotations. Material in regular type represents the author's comments.)

(The following paragraph is not part of a trial transcript, but is quoted from the website Voices For Evolution)

In 1990, in Webster v. New Lenox School District, the Seventh Circuit Court of Appeals found that a school district may prohibit a teacher from teaching creation science, in fulfilling its responsibility to ensure that the First Amendment's establishment clause is not violated, and religious beliefs are not injected into the public school curriculum. The court upheld a district court finding that the school district had not violated Webster's free speech rights when it prohibited him from teaching "creation science," since it is a form of religious advocacy. (Webster v. New Lenox School District #122, 917 F. 2d 1004)

This case stands out among the others because it does not deal directly with Science teaching. The plaintiff in this case was a junior-high Social Studies teacher. As shown in the background section of the transcript, Webster's classroom textbook contained a statement that the world was more than four billion years old. It didn't propose it as a theory. It didn't say there was a possibility it might not be four billion years old. It just made a statement without qualification.

The one thing that ought to be noticed here is, in the case of Segrave vs State of California, 1981, that state had an anti-dogmatism law. This should mean, in simplest terms, that theories can't be taught as facts. Yet if you look at the thousands of textbooks and millions of even quasi-educational materials on every possible subject available to children, from post-graduate doctoral material to preschool "world explorer" cartoons, this is the way they always state their dogma. Not "studies suggest an age of four billion years for the earth," or "Scientists have said the earth may be more than four billion years old." It's stated as a fact. Our eyes run right past it now because it's so common. History books begin with Cro-Magnon and Neanderthal man, or maybe even Australopithecus Afarensis, the so-called "Lucy" fossil. Literature survey books cover "prehistory" such as cave paintings and attribute them to an earlier link to modern man. And the information is presented as factual, not theoretical.

Author's Note: The format of this transcript is somewhat unusual so the following should help to clarify its presentation here. The transcript is presented in its original format. Structure and headings (such as the Roman numeral I and the heading entitled Background) are part of the original document. All quoted material appears in italic type. Commentary by the author appears in regular type, in various places throughout. Occasionally the author picks up a quote to repeat, and this material also appears in italic type. The heading information from the very beginning of the original document, with the title of the case and various legal listings has been omitted. The transcript reproduced here begins with the name of the presiding judge and a statement of the case.

Ripple, Circuit Judge. Ray Webster sought injunctive and declaratory relief based on his claim that the New Lenox School District violated his first and fourteenth amendment rights by prohibiting him from teaching a nonevolutionary theory of creation in the classroom. He appeals the dismissal of his complaint for failure to state a claim. For the following reasons, we affirm the judgment of the district court.

I

Background

The district court dismissed Mr. Webster's suit for failure to state a claim upon which relief can be granted. See Fed. R. Civ. P. 12(b)(6). The grant of a motion to dismiss is, of course, reviewed de novo. Villegas v. Princeton Farms, Inc., 893 F.2d 919, 924 (7th Cir. 1990); Corcoran v. Chicago Park Dist., 875 F2d 609, 611 (7th Cir. 1989). It is well settled that, when reviewing the grant of a motion to dismiss, we must assume the truth of all well-pleaded factual allegations and make all possible inferences in favor of the plaintiff. Janowsky v. United States, No. 89-2219, slip op. at 4 (7th Cir. Sept. 17, 1990); Rogers v. United States, 902, F.2d 1268, 1269 (7th Cir. 1990).

A complaint should not be dismissed "unless it appears beyond doubt that the plaintiff can prove no set of facts in support of his claim which would entitle him to relief." Conley v. Gibson, 355 U. S. 41, 45-46 (1957). This obligation is especially serious when, as here, we deal with allegations involving the freedom of expression protected by the first amendment. See Stewart v. District of Columbia Armory Bd., 863 F.2d 1013, 1017-18 (D.C. Cir. 1988) ("where government action is challenged on first amendment grounds, a court should be especially 'unwilling to decide the legal questions posed by the parties without a more thoroughly developed record of proceedings in which the parties have an opportunity to prove those disputed factual assertions upon which they rely'") (quoting City of

Los Angeles v. Preferred Communications, 476 U.S. 488, 494 (1986)). Courts may, however, consider exhibits attached to the complaint as part of the pleadings. Beam v. IPCO Corp., 838 F.2d 242, 244 (1988). With these constraints in mind, we set forth the pertinent facts.

A. Facts

Ray Webster teaches social studies at the Oster-Oakview Junior High School in New Lenox, Illinois. In the spring of 1987, a student in Mr. Webster's social studies class complained that Mr. Webster's teaching methods violated principles of separation between church and state. In addition to the student, both the American Civil Liberties Union and the Americans United for the Separation of Church and State objected to Mr. Webster's teaching practices. Mr. Webster denied the allegations. On July 31, 1987, the New Lenox school board (school board), through its superintendent, advised Mr. Webster by letter that he should restrict his classroom instruction to the curriculum and refrain from advocating a particular religious viewpoint.

Believing the superintendent's letter vague, Mr. Webster asked for further clarification in a letter dated September 4, 1987. In this letter, Mr. Webster also set forth his teaching methods and philosophy. Mr. Webster stated that the discussion of religious issues in his class was only for the purpose of developing an open mind in his students. For example, Mr. Webster explained that he taught nonevolutionary theories of creation to rebut a statement in the social studies textbook indicating that the world is over four billion years old.

Therefore, his teaching methods in no way violated the doctrine of separation between church and state. Mr. Webster contended that, at most, he encouraged students to explore alternative viewpoints.

The superintendent responded to Mr. Webster's letter on October 13, 1987. The superintendent reiterated that advocacy of a Christian viewpoint was prohibited, although Mr. Webster could discuss objectively the historical relationship between church and state when such discussions were an appropriate part of the curriculum. Mr. Webster was specifically instructed not to teach creation science, because the teaching of this theory had been held by the federal courts to be religious advocacy.

Mr. Webster brought suit, principally arguing that the school board's prohibitions constituted censorship in violation of the first and fourteenth amendments. In particular, Mr. Webster argued that the school board should permit him to teach a nonevolutionary theory of creation in his social studies class.

Unfortunately the actual correspondence that preceded this case is not available, only summaries which in themselves seem biased in favor the school district. It would also be interesting to know exactly what the student's original complaint was and how all of this got started. It would seem necessary to know how Webster worded his letters or even how he presented his material in the classroom. But these facts are not deemed important enough to be included, apparently. We only know that Webster said he only wished to present a "nonevolutionary" theory (not a fact, not a religious belief, only a theory) to balance the statement in the textbook, which was not presented as a theory, but as a fact. The students would logically assume their textbook taught them facts, unless it told them it was presenting a theory.

B. The District Court

The district court concluded that Mr. Webster did not have a first amendment right to teach creation science in a public school. The district court began by noting that, in deciding whether to grant the school district's motion to dismiss, the court was entitled to consider the letters between the superintendent and Mr. Webster because Mr. Webster had attached these letters to his complaint as exhibits. In particular, the district court determined that the October 13, 1987 letter was critical; this letter clearly indicated exactly what conduct the school district sought to proscribe. Specifically, the October 13 letter directed that Mr. Webster was prohibited from teaching creation science and was admonished not to engage in religious advocacy. Furthermore, the superintendent's letter explicitly stated that Mr. Webster could discuss objectively the historical relationship between church and state.

The case here rests on whether Webster was denied the freedom to teach what he as an educator thought was right for his students to know. He thought it was right for his students to know that determining the age of the earth is theoretical. He taught them that there was more than one theory used to determine this information and that more than one conclusion could be reached. That is his statement of what he did and why he did it. Unless he committed perjury and lied about what he taught, it is necessary to accept his statement as fact.

In order to explain how it is possible that the factual statement in the textbook might not in fact be a fact, it was necessary to bring up an alternate theory for determining the age of the earth. It should be significant to note that it was apparently not sufficient for Webster to say, "this statement about the age of the earth is part of a theory, the theory of evolution. There are other theories about how to determine how old the earth is, but you already know all about them, so we'll just leave it at that."

He was unable to do that, because it's extremely likely that the students in his classroom had no idea that statement wasn't a fact or what any other theories that challenge it might be. We are not privy to the discussion in the classroom. It wasn't drawn out in interviews and made part of the case writings. It is possible, however, to speculate based on the modern teaching texts and quasi-educational materials available that students are not being given more than one theory to consider. They are being presented with statements of fact about things that are in reality still theories. It was therefore necessary for Webster to spell out another theory that encompasses how to determine the age of the earth. It was necessary to point out that the textbook had a fact that wasn't a fact. It was then necessary to explain why it wasn't a fact. He was a teacher. It was his job to tell students things they didn't know, even if those things weren't part of the curriculum.

The letter Webster got from the school district, the one considered critical to the disposition of the case, tells him he was prohibited from teaching creation science and was admonished not to engage in religious advocacy. Furthermore, the superintendent's letter explicitly stated that Mr. Webster could discuss objectively the historical relationship between church and state. Propping itself up on the crutch of previous court decisions, the school district considered teaching creation science teaching religion. Apparently it was not just teaching religion, but advocating it. It must not be possible to teach a concept without advocating it. Webster presented creation science as a concept in his classroom. Therefore he advocated it. Therefore he must be proscribed from it. Of course he was permitted to deal with historic church and state relationships. He was permitted to teach something, even though it really had nothing to do with correcting the error in his textbook. Because he was allowed to teach something about religion, his free speech wasn't infringed upon. Case dismissed.

The district court noted that a school board generally has wide latitude in setting the curriculum, provided the school board remains within the boundaries established by the constitution. Because the establishment clause prohibits the enactment of any law "respecting an establishment of religion," the school board could not enact a curriculum that would inject religion into the public schools. U.S. Const. amend. I. Moreover, the district court determined that the school board had the responsibility to ensure that the establishment clause was not violated.

The district court then framed the issue as whether Mr. Webster had the right to teach creation science. Relying on Edwards v. Aguillard, 482 U.S. 578 (1987), the district court determined that teaching creation science would constitute religious advocacy in violation of the first amendment and that the school board correctly prohibited Mr. Webster from teaching such material. The court further noted:

Webster has not been prohibited from teaching any nonevolutionary theories or from teaching anything regarding the historical relationship between church and state. Martino's [the superintendent] letter of October 13, 1987 makes it clear that the religious advocacy of Webster's teaching is prohibited and nothing else. Since no other constraints were placed on Webster's teaching, he had no basis for his complaint and it must fail.

Webster v. New Lenox School Dist., Mem. op, at 4-5 (N.D. Ill. May 25 1989). Accordingly, the district court dismissed the complaint.

II

Analysis

At the outset, we note that a narrow issue confronts us: Mr. Webster asserts that he has a first amendment right to determine the curriculum content of his junior high school class. He does not, however, contest the general authority of the school board, acting through its executive agent, the superintendent, to set the curriculum.

This case does not present a novel issue. We have already confirmed the right of those authorities charged by state law with curriculum development to require the obedience of subordinate employees, including the classroom teacher. Judge Wood expressed the controlling principle succinctly in Palmer v. Board of Educ., 603 F.2d 1271, 1274 (7th Cir. 1979), cert. denied, 444 U.S. 1026 (1980), when he wrote:

Parents have a vital interest in what their children are taught. Their representatives have in general prescribed a curriculum. There is a compelling state interest in the choice and adherence to a suitable curriculum for the benefit of our young citizens and society. It cannot be left to individual teachers to teach what they please.

Yet Mr. Webster, in effect, argues that the school board must permit him to teach what he pleases. The first amendment is "not a teacher license for uncontrolled expression at variance with established curricular content." Id. at 1273. See also Clard v. Holmes, 474 F.2d 928 (7th Cir.) (holding that individual teacher has no constitutional prerogative to override the judgment of his superiors as to proper course content), cert. denied, 411 U.S. 972 (1973). Clearly, the school board had the authority and the responsibility to ensure that Mr. Webster did not stray from the established curriculum by injecting religious advocacy into the classroom. "Families entrust public schools with the education of their children, but condition their trust on the understanding that the classroom will not purposely be used to advance religious views that may conflict with the private beliefs of the student and his or her family." Edwards v. Aguillard, 482 U.S. 578, 584 (1987).

There is no indication that Webster said he had a right to determine the curriculum content of his junior high school class, nor that Mr. Webster, in effect, argues that the school board must permit him to teach what he pleases. (already cited) The court acknowledges that he accepted the authority of the board. He had the required textbook and was teaching from it, or he wouldn't have run across the passage in question. He would have saved himself from a great deal of trouble if he had simply not assigned his class to read that part of the book. Yet Webster was not interested in stifling even what he considered to be a lie, or at least not a fact. He presented the statement, and then he offered an alternative. It seems extreme to state that he was guilty of "uncontrolled expression at variance with established curricular content." He also was not guilty of betraying the trust of the School District families by abusing his position to advance conflicting religious views. The key word is "advancing." It implies putting something forward as superior, similar to the idea of "advocacy." A teacher presents a concept in his class, and the board and the court assumes he is in favor of it, that he wants his students to believe it, not just listen to it. But it seems that is only the case if the teacher presents Creation Science. He can objectively present anything else. Not that.

A teacher may say, "the 'Final Solution' for dealing with the Jews was to put them in concentration camps and kill them." Is the teacher advocating death camps for Jews? Of course not. He is presenting to his class a historical position. Unless the teacher is insane, his position is quite opposite from advocating what he states. Surely other "solutions" were possible for Hitler (granting that there was a problem of the nature under discussion, which, again, no sane person would agree with) and may even have been presented to Hitler at the time of this historical event. But Hitler wanted this one. Only this one. He may even have proscribed his advisors from bringing any other theories up. He certainly would not have been interested in possible solutions that appeared to advocate or advance a religious position.

If we ever would have wanted a government to permit the presentation of an alternate theory to deal with an issue, even a religious-based theory, that would have been the time. Alternate theories were apparently stifled, because, ironically enough, survival of the fittest, one of Evolution's foundational tenets, one upon which Hitler based many of his actions, couldn't survive on its own. It had to be protected then, and it has to be protected by the American justice system now.

A junior high school student's immature stage of intellectual development imposes a heightened responsibility upon the school board to control the curriculum. See Zykan v. Warsaw Community School Corp., 631 F.2d 1300, 1304 (7th Cir. 1980). We have noted that secondary school teachers occupy a unique position for influencing secondary school students, thus creating a concomitant power in school authorities to choose the teachers and regulate their pedagogical methods. Id. "The state exerts great authority and coercive power through mandatory attendance requirements, and because of the students' emulation of teachers as role models and the children's susceptibility to peer pressure." Edwards, 482 U.S. at 584 (footnote omitted).

It is true that the discretion lodged in school boards is not completely unfettered. For example, school boards may not fire teachers for random classroom comments. Zykan, 631 F.2d at 1305. Moreover, school boards may not require instruction in a religiously inspired dogma to the exclusion of other points of view. Epperson v. Arkansas, 393 U.S. 97, 106 (1968).

Interesting that Epperson v. Arkansas is brought up to support exactly the opposite of what it accomplished. Creation Science, another point of view, was squashed and the religious dogma of Evolution was exclusively preferred.

This complaint contains no allegation that school authorities have imposed “a pall of orthodoxy” on the offerings of the entire public school curriculum, Keyishian v. Board of Regents, 385 U.S. 589, 603 (1967),” which might either implicate the state in the propagation of an identifiable religious creed or otherwise impair permanently the student’s ability to investigate matters that arise in the natural course of intellectual inquiry.” Zykan, 631 F2d at 1306. Therefore, this case does not present the issue of whether, or under what circumstances, a school board may completely eliminate material from the curriculum. Cf. Zykan, 631 F.2d at 1305-06 (school may not flatly prohibit teachers from mentioning relevant material). Rather, the principle that an individual teacher has no right to ignore the directives of duly appointed education authorities is dispositive of this case. Today, we decide only that, given the allegations of the complaint, the school board has successfully navigated the narrow channel between impairing intellectual inquire and propagating a religious creed.

The wording of the previous opinion implies that the school board must be guilty of promoting a religion throughout the entire curriculum before it can be said to be in violation of the separation principle or infringing upon an individual teacher's free expression right. A teacher, however, can be denied the right to express one idea, apparently. Never mind that the entire curriculum required by the state and the school board is riddled with evolutionary language. This case is a perfect example of that. Who would have thought evolution would be an issue in a social studies class?

This is a concept appearing absolutely everywhere in curriculum. It can't be repeated often enough that it exists in every subject, often in incidental but purposeful references, as a fact, not as a theory. The state and the school boards absolutely are guilty of promoting this dogma exclusive of any other, in every subject. Try to find a state-approved textbook in any academic subject that does not bring up the subject of billions of years of earth age, or discuss man as being at first entirely primitive and gradually becoming civilized, or note that the extinction of certain types of animals or plants also indicates that more adaptable types replaced them. These are evolutionary concepts.

Here, the superintendent concluded that the subject matter taught by Mr. Webster created serious establishment clause concerns. Cf. Edwards, 482 U.S. at 583-84 ("The Court has been particularly vigilant in monitoring compliance with the Establishment Clause in elementary and secondary schools."); Epperson, 393 U.S at 106 (school may not adopt programs that aid or oppose any religion). As the district court noted, the superintendent's letter is directed to this concern. "[E]ducators do not offend the First Amendment so long as their actions are reasonably related to legitimate pedagogical concerns." Hazelwood School Dist. v. Kuhlmier, 484 U.S 260, 278 (1988). Given the school board's important pedagogical interest in establishing the curriculum and legitimate concern with possible establishment clause violations, the school board's prohibition on the teaching of creation science to junior high students was appropriate. See Palmer v. Board of Educ., 603 F.2d 1274 (7th Cir. 1979) (school board has "compelling" interest in setting the curriculum). Accordingly, the district court properly dismissed Mr. Webster's complaint.

The school in this case adopted a program that both aided one religion and opposed another. The dogma of evolution was once again aided, and the truth was opposed.

Peloza v. Capistrano School District, 1994

(Author's Note: Italicized material represents direct quotations. Material in regular type represents the author's comments. This section also has material from the website Vine & Fig Tree.)

(The following paragraph is not part of a trial transcript, but is quoted from the website Voices For Evolution)

In 1994, in Peloza v. Capistrano School District, the Ninth Circuit Court of Appeals upheld a district court finding that a teacher's First Amendment right to free exercise of religion is not violated by a school district's requirement that evolution be taught in biology classes. Rejecting plaintiff Peloza's definition of a "religion" of "evolutionism", the Court found that the district had simply and appropriately required a science teacher to teach a scientific theory in biology class. (John E. Peloza v. Capistrano Unified School District, (1994) 917 F. 2d 1004)

A summary of the case follows, taken from the transcript appearing on the website *www.talkorigins.org*.

SUMMARY

High school biology teacher brought action against school district, its board of trustees, and various personnel at high school, challenging school district's requirement that he teach evolutionism, as well as school district order barring him from discussing his religious beliefs with students. The United States District Court, Central District of California, David W. Williams, J., 782 F.Supp. 1412, dismissed and awarded attorney fees to school district. Teacher appealed. The Court of Appeals held that: (1) teacher failed to state claim for violation of establishment clause of First Amendment in connection with school district's requiring him to teach evolution, i.e., that higher life forms evolved from lower ones; (2) school district's restriction on teacher's right of free speech in prohibiting teacher from talking with students about religion during school day, including times when he was not actually teaching class, was justified by school district's interest in avoiding establishment clause violation; (3) teacher's allegations of injury to his reputation as result of allegedly defamatory statements made to and about him were insufficient to support claim for deprivation of liberty interest under § 1983; but (4) teacher's complaint was not entirely frivolous, precluding award of costs and attorney fees under Rule 11 and § 1988.

The case disposition almost entirely ruled against every point the teacher Peloza brought up. One judge, Pole, offered a partial dissenting opinion, of which an excerpt follows.

I am in agreement with the majority's resolution of John Peloza's Establishment Clause and Due Process Clause claims. However, because I believe we can dismiss Peloza's free speech claims only by turning a deaf ear to the procedural posture of this case, I respectfully dissent from parts I. B and II of the majority opinion.

I

Schoolteacher John Peloza seeks a declaratory judgment permitting him to "respond to student-initiated inquiries ... regarding religion" during contract time. The majority opinion concludes that if Peloza's discussions would constitute an establishment of religion, the District may permissibly limit those discussions, even though such limitations restrict Peloza's free speech. With this I have no quarrel. But the majority's premise is that any discussions Peloza might have do constitute such an establishment, and I am unpersuaded that we may reach such a conclusion in the case's present posture.

This is an appeal from the granting of a Rule 12(b)(6) motion. As such, we are not permitted to affirm dismissal of the complaint "unless it appears beyond doubt that plaintiff can prove no set of facts in support of his claim which would entitle him to relief." Love V. United States, 915 F.2d 1242, 1245 (9th Cir.1989). At this stage, we know almost nothing about what past or future discussions might involve. I can imagine a wide range of circumstances and questions "regarding religion" which Peloza could permissibly answer without violating the Establishment Clause. For example, a student might come to a teacher during lunch and ask about Malcolm X or Martin Luther King's religious beliefs, and how and why they evolved, or about the origins of Islam, or what the seven great religions of the world were. Such questions would certainly be "regarding religion," student-initiated, and during contract time. As such, they fall within the class of discussions Peloza seeks to be permitted, yet it is hard to see how the descriptive role a teacher would have in responding to these questions would work any violation of the Establishment Clause.

The majority holding only makes sense if we presume that we know what kinds of questions are being asked and what kinds of answers Peloza would give. In the posture of this case, where we must reverse if there are any facts Peloza could conceivably prove which would entitle him to relief, this is a presumption we are forbidden from making. As a result, the majority holding means that any response to a student-initiated inquiry "regarding religion" during contract time, other than "Ask someone else," works a violation of the

Establishment Clause. I cannot join in such a broad legal holding, and indeed the case law forbids it:

In each case, the inquiry calls for line-drawing; no fixed, per se rule can be framed. The Establishment Clause like the Due Process Clauses is not a precise, detailed provision in a legal code capable of ready application.... The line between permissible relationships and those barred by the Clause can no more be straight and unwavering than due process can be defined in a single stroke or phrase or test. The Clause erects a "blurred, indistinct, and variable barrier depending on all the circumstances of a particular relationship." Lemon V. Kurtzman, 403 U.S. [602, 614, 91 S.Ct. 2105, 2112, 29 L.Ed.2d 745 (1971)]. Lynch V. Donnelly, 465 U.S. 668, 678-79, 104 S.Ct. 1355,1362, 79 L.Ed.2d 604 (1984).

Roberts V. Madigan, 921 F.2d 1047 (10th Cir.1990), upon which the majority relies, is not to the contrary. There, the court had before it a host of particulars: the conduct at issue involved a teacher displaying religious books and a poster reading "You have only to open your eyes to see the hand of God" in the classroom. Id. at 1049. That court also had the benefit of a district court factual determination that the conduct "created the appearance that [the teacher] was seeking to advance his religious views." Id. As this case stands, we know far less.

The majority impermissibly attempts to narrow the scope of Peloza's complaint by relying on a written warning from the school district which Peloza has incorporated into the complaint. The letter forbids Peloza from "attempt[ing] to convert students to Christianity or initiating conversations about your religious beliefs." Complaint at 45. Were this all that the complaint said, I would have little trouble joining the majority. But the complaint alleges more; it contends that "the school district ... has directed Plaintiff not to discuss any religious matters during any of this 'instructional time,' including student-initiated conversations regarding religion during lunch, class breaks, and before and after school hours." Complaint at 3. This allegation we must take as true. If all that lies behind it is the far narrower warning the majority cites, then Peloza's case will not be long for this world. But we may not presume that this is so.

I believe that, in a broad range of cases, the majority and I could agree about what would or would not constitute a violation of the Establishment Clause. But the majority errs in presuming to know that what is at stake here is Peloza's right to "discuss his religious beliefs" with students. In doing so, it ignores the fact that this is a Rule 12(b)(6) case. More generally, it gives short shrift to the possibility that we may well be limiting free speech more broadly than the state's compelling interest in avoiding an establishment of religion would warrant.

II

I join in the majority's part II insofar as it dismisses Peloza's § 1985(3) due process and Establishment Clause claims based on his failure to properly allege a violation of these rights. However, because I conclude that Peloza's free speech claim should not have been dismissed, I would also remand, rather than dismiss, his 1985(3) claim based on alleged free speech violations.

III

Religion has been used to justify the suppression of speech for centuries. See Everson V. Board of Ed,, 330 U.S. 1, 8-10, 67 S.Ct. 504, 5074)9, 91 L.Ed. 711 (1947). With the development of a vigorous First Amendment jurisprudence, we have quelled some of the worst abuses. But points of tension remain. We must thus remain vigilant to ensure that in our rush to preserve certain fundamental rights, we do not trample others. Caution is of the essence; only through a methodical and fact-specific jurisprudence can we hope to achieve a proper accommodation.

For the reasons stated above, I respectfully dissent.

1. On appeal, Peloza abandoned his equal protection argument.

2. The Establishment Clause of the First Amendment provides that "Congress shall make no law respecting an establishment of religion..." The Fourteenth Amendment incorporates the Establishment Clause's prohibitions against offending state action as well. Board of Education v. Pico, 457 U.S. 853, 864, 102 S.Ct.2799, 2806-07, 73 L.Ed.2d 435 (1982).

3. See Webster's Third New Int'l Dictionary (G. & C. Merriam Co. Springfield, MA. 1969). p.789 ("evolutionism: 1: a theory of evolution (as in philosophy, biology, or sociology) - See Darwinism 2: adherence to or belief in evolution esp. of living beings").

4. According to Webster's, religion is the "belief in and reverence for a supernatural power accepted as

the creator and governor of the universe." Webster's II New Riverside University Dictionary 993.

5. See Smith v. Board of School Com'rs of Mobile County, 827 F.2d 684, 690-95 (11th Cir.1987) (refusing to adopt district court's holding that "secular humanism" is a religion for Establishment Clause purposes; deciding case on other grounds); United States v. Allen, 760 F.2d 447, 450-51 (2d Cir.1985) (quoting Tribe, American Constitutional Law 827-28 (1987), for the proposition that, while "religion" should be broadly interpreted for Free Exercise Clause purposes, "anything 'arguably non-religious' should not be considered religious in applying the establishment clause").

6. The dissent claims this interpretation impermissibly narrows the scope of Peloza's complaint. However, the very sentence quoted by the dissent, Dissent at p.12064, focuses not on the definition of "religious matters," but on the definition of "instructional time." We agree with the dissent that a complaint must be read charitably at the Rule 12(b)(6) stage. However, a reviewing court need not go so far as to invent claims not within the reasonable intendment of the complaint.

7. As with his equal protection claim under section 1983, Peloza appears to have dropped his equal protection claim from his appeal to this court.

8. See United Brotherhood of Carpenters & Joiners of America, Local 610, AFL-CIO v. Scott, 463 U.S. 825, 830, 103 S.Ct. 3352,3357, 77 L.Ed.2d 1049 (1983) (hate speech rights protected by section 1985 so long as the State is involved in the conspiracy alleged). As to due process rights, there appears to be some confusion within this circuit. Older cases

have stated that section 1985(3) provides no remedy for violation of due process rights, Cohen V. Norris, 300 F.2d 24, 28 (9th Cir. 1962) (dicta); Mitchell V. Greenough, 100 F.2d 184, 187 (9th Cir.1938) (holding), cert. denied. 306 U.S. 659, 59 S.Ct. 788, 83 L.Ed. 1056 (1939). In some more recent cases, we have allowed claims of due process violations to proceed under section 1985(3) without comment. See Judie V. Hamilton. 872 F.2d 919,924 (9th Cir.1989); Padway V. Palches. 665 F.2d 965, 969 (9th Cir. 1982). See Taylor V. Gilmartin, 686 F.2d 1346, 1358 (10th Cir.1982) (First Amendment freedom of religion protected by section 1985(3)). cert. denied, 459 U.S. 1147, 103 S.Ct. 788,74 L.Ed.2d 994 and cert. denied, 463 U.S. 1229, 103 S.Ct. 3570, 77 L.Ed.2d 1411 (1983); Action V. Gannon. 450 F.2d 1227, 1234 (8th Cir.1971) (satne); Cooper V. Molko, 512 F.Supp. 563, 570 (N.D.Cal.1981) (same); but see Africa V. Anderson, 510 F.Supp. 28, 30 (E.D.Penn.1980) (freedom of religion not protected by section 1985(3)).

The dissenting judge touches on the key point of the whole argument all of these evolution-related court case have shared. Opposition to the teaching of Evolution can only be on religious grounds and all religious talk must be stifled. Only in that way can Evolution opposition be stifled.

The judge properly questioned how a school district could make a blanket order forbidding the teacher such a wide range of activities. First, they prohibited any discussion of religion at all. Second, they forbade conversations outside the classroom and class time. Third, they forbade even student-initiated conversations.

With the development of a vigorous First Amendment jurisprudence, we have quelled some of the worst abuses. But points of tension remain. We must thus remain vigilant to ensure that in our rush to preserve certain fundamental rights, we do not trample others. Caution is of the essence; only through a methodical and fact-specific jurisprudence can we hope to achieve a proper accommodation.

This is a well-spoken statement, but it is too little, too late. The damage is done. The fundamental right to teach truth and reject error has been trampled repeatedly by court decisions designed solely to protect the State religion of Secular Humanism and its Bible of Evolution Dogma. For a judge to stop now and think of how the judicial system might have gone too far is sad. To date, judges have fervently served as the High Priests of Humanism and defended the dogma against all comers.

In an earlier court case Justice Black was quoted as having brought up Thomas Jefferson's "wall of separation" concept. That non-constitutional statement has since been solidly welded into place and given rise to a host of misinterpretations and abuses of intent and power.

A later case chose to quote Robert Frost's "Mending Wall," as if that were proof of the need for separation when its point is really the opposite. Is it possible that judges do not at all understand the intention of the founding fathers in the Constitution and documents of the same time period? And is it possible that even the United States Supreme Court had a different interpretation of the "separation principle" long after the days of the founding (see the 1844 quotation below) but before Secular Humanism became the law of the land?

The following comes from the website Vine & Fig Tree, which describes itself as Supporting: love, joy, peace, patience, gentleness, goodness, faith, meekness, sobriety [and] Opposing: Secularism, Humanism, Anti-Family Sex, Hedonism, Autonomy, Totalitarianism, and Mass Death.

In stark contrast to the myth of separation, the Founders believed that schools should positively and affirmatively teach religion. Every single person who signed the Constitution believed that religious and moral inculcation was the purpose of schools. Peloza is light-years away from the original intent of the Constitution. Consider Samuel Adams:

As piety, religion, and morality have a happy influence on the minds of men, in their public as well as private transactions, you will not think it unseasonable, although I have frequently done it, to bring to your remembrance the great importance of encouraging our University, town schools, and other seminaries of education, that our children and youth while they are engaged in the pursuit of useful science, may have their minds impressed with a strong sense of the duties they owe to God. If we continue to be a happy people, that happiness must be assured by the enacting and executing of the reasonable and wise laws expressed in the plainest language and by establishing such modes of education as tend to inculcate in the minds of youth the feelings and habits of "piety, religion and morality." (Addressing the Legislature of Mass., 1/16/1795)

Let divines and philosophers, statesmen and patriots, unite their endeavors to renovate the age, by impressing the minds of men with the importance of educating their little boys and girls, of inculcating in the minds of youth the fear and love of the Deity. . . and, in subordination to these great principles, the love of their country. . . . In short, of leading them in the study and practice of the exalted virtues of the Christian system. Letter to John Adams, 1790, who wrote back: "You and I agree." Four Letters: Being an Interesting Correspondence Between Those Eminently Distinguished Characters, John Adams, Late President of the United States; and Samuel Adams, Late Governor of Massachusetts. On the Important Subject of Government (Boston: Adams and Rhoades, 1802) pp. 9-10

It has been observed that "education has a greater influence on manners than human laws can have." [A] virtuous education is calculated to reach and influence the heart and to prevent crimes. . . . Such an education, which leads the youth beyond mere outside show, will impress their minds with a profound reverence of the Deity [and] . . . will excite in them a just regard to Divine revelation. The Life and Public Services of Samuel Adams, Wm.Wells., ed. (Boston: Little, Brown, & Co., 1865) Vol.III, p. 327.

Art. 3. Religion, morality, and knowledge, being necessary to good government and the happiness of mankind, schools and the means of education shall forever be encouraged. Northwest Ordinance, 1787

In my view, the Christian religion is the most important and one of the first things in which all children, under a free government, ought to be instructed. . . . No truth is more evident to my mind than that the Christian religion must be the basis of any government intended to secure the rights and privileges of a free people. The opinion that human reason left without the constant control of Divine laws and commands will preserve a just administration, secure freedom and other rights, restrain men from violations of laws and constitutions, and give duration to a popular government is as chimerical as the most extravagant ideas that enter the head of a maniac . . . . Where will you find any code of laws among civilized men in which the commands and prohibitions are not founded on Christian principles? I need not specify the prohibition of murder, robbery, theft [and] trespass. Noah Webster, Letters, Harry A Warfel, ed., (NY: Library Publishers, 1953) pp. 453-454, to David McClure, Oct. 25, 1836.

Thomas Jefferson's good friend Benjamin Rush, after he signed the Declaration of Independence, was the first Founding Father to call for free public schools. He said:

[T]he only foundation for a useful education in a republic is to be laid in religion. Without this there can be no virtue, and without virtue there can be no liberty, and liberty is the object and life of all republican governments. Without religion, I believe that learning does real mischief to the morals and principles of mankind.(Benjamin Rush, Essays, Literary, Moral, and Philosophical, 1798, p.6 ["On the Mode of Education Proper in a Republic"])

Rush was clearly a Christian, but no "fundamentalist nut." In his paper entitled, "A

Defense of the Use of the Bible as a Schoolbook," Rush argued,

[T]he only means of establishing and perpetuating our republican forms of government . . . is the universal education of our youth in the principles of Christianity by means of the Bible. For this Divine book, above all others, favors that equality among mankind, that respect for just laws, and those sober and frugal virtues, which constitute the soul of republicanism.

Daniel Webster reflected the views of every single Signer of the Constitution:

We regard it [public instruction] as a wise and liberal system of police by which property and life and the peace of society are secured. We seek to prevent in some measure the extension of the penal code by inspiring a salutary and conservative principle of virtue and of knowledge. [1]

[However, t]he attainment of knowledge does not comprise all which is contained in the larger term of education. The feelings are to be disciplined; the passions are to be restrained; true and worthy motives are to be inspired; a profound religious feeling is to be instilled, and pure morality inculcated. [Four years later, the U.S. Supreme Court would agree that this could only be done by having the government teach the Bible.] [2]

The cultivation of the religious sentiment represses licentiousness . . . inspires respect for law and order, and gives strength to the whole social fabric.[3]

[1] Works of Daniel Webster (Boston: Little, Brown, and Co., 1853) vol I, pp 41-42, Dec 22., 1820.[2] vol

II, pp 107-108, Oct 5: 1840[3] vol II, p 615, July 4, 1851

The Father of his Country warned:

And let us with caution indulge the supposition that morality can be maintained without religion. Whatever may be conceded to the influence of refined education on minds of peculiar structure, reason and experience both forbid us to expect that national morality can prevail in exclusion of religious principle And secularists would be quick to point out that Washington was less Biblically-oriented than the most influential educators in the nation, such as Benjamin Rush and Noah Webster.

All the scholars are required to live a religious and blameless life according to the rules of God's Word, diligently reading the holy Scriptures, that fountain of Divine light and truth, and constantly attending all the duties of religion . . . . All the scholars are obliged to attend Divine worship in the College Chapel on the Lord's Day and on Days of Fasting and Thanksgiving appointed by public Authority. The Laws of Yale College in New Haven in Connecticut (New Haven: Josiah Meigs, 1787) p. 5-6, ch II, art. 1,4

William Samuel Johnson, signer of the Constitution, was appointed Columbia's first president. Under him, It is expected that all students attend public worship on Sundays. Columbia Rules (NY: Samuel Loudon, 1785) 5-8

Johnson's views on public education were similar to those of every other signer of the Constitution. In his commencement address, he told the graduates:

You this day, gentlemen, . . . have . . . received a public education, the purpose whereof hath been to qualify you the better to serve your Creator and your country . . . . Your first great duties, you are sensible, are those you owe to Heaven, to your Creator and Redeemer. Let these be ever present to your minds, and exemplified in your lives and conduct. Imprint deep upon your minds the principles of piety towards God and a reverence and fear of His holy name. The fear of God is the beginning of wisdom [Proverbs 9:10]. Remember too, that you are the redeemed of the Lord, that you are bought with a price, even the inestimable price of the precious blood of the Son of God. . . . Love, fear and serve Him as your Creator, Redeemer, and Sanctifier. Acquaint yourselves with Him in His Word and holy ordinances. Make Him your friend and protector and your felicity is secured both here and hereafter.

Early US Supreme Court decisions agreed that in a Christian nation such as America, the Bible must be taught in all government-run schools.

In 1844, the Court was asked, Can the state enforce a will which creates a government-operated school which will not teach the Bible? The Supreme Court said that the very idea of a school which will not teach the Bible is contrary to the legal foundations of this Christian nation.

It is unnecessary for us, however, to consider what would be the legal effect of a devise in Pennsylvania for the establishment of a school or college, for the propagation of . . . Deism, or any other form of infidelity. Such a case is not to be presumed to exist in a Christian country; and therefore it must be made out by clear and indisputable proof.

The government made firm assurances that the Bible would be taught in the school, and the will was approved. (Vidal v. Girard's Executors)

The Vidal Court, as it talks about Christianity and the Bible, sounds more like David Barton than anything one would hear from the post-1947 Court. The Vidal Court said that the government in its school "may, nay must impart to their youthful pupils . . . the Bible, and especially the New Testament," which must "be read and taught as a divine revelation in the college -- its general precepts expounded, its evidences explained, and its glorious principles of morality inculcated." The Court asked rhetorically:

Where can the purest principles of morality be learned so clearly or so perfectly as from the New Testament? Where are benevolence, the love of truth, sobriety, and industry, so powerfully and irresistibly inculcated as in the sacred volume?

The Bible MUST be taught in government schools, the 1844 US Supreme Court declared.

You would NEVER EVER hear language like this from the modern secularist Court. But you ALWAYS heard language like this from the Founding Fathers.

The "separation of church and state" is a myth.

Appendix Three: Supplementary Material for Sections Two and Three

Higher Criticism

For the prophecy came not in old time by the will of man: but holy men of God spake as they were moved by the Holy Ghost. (2 Peter 1:21).

All Scripture is given by inspiration of God, and is profitable for doctrine, for reproof, for correction, for instruction in righteousness. (2 Timothy 3:16).

Until well after the American Civil War, biblical scholars in the United States accepted these verses as truth. Everyone, conservative or liberal, acknowledges that human beings wrote down the words of the Scriptures. After that admission, two positions are possible to answer the question of who wrote the Bible. Either it was written by the people who claim to have written it, or it was written much later, by unknown authors or editors. If it was written much later, by unknown authors or editors, it does not contain truth, but is a patchwork of mythologies and lies and not worthy of serious study.

The Bible itself makes claims as to its authorship. For example, Moses wrote the first five books, with later minor additions by Joshua and Samuel. Ezra performed major editorial work on most of the Old Testament to bring it into its present form. Samuel and the School of the Prophets are responsible for the books through II Samuel, Ezra was responsible for Chronicles, Kings, and the book bearing his name, and many other books bear the names of their authors, continuing into the New Testament with the gospels and General Epistles. There is no evidence of an editor of the New Testament with the possible exception of the book of Hebrews. Works acknowledged to be written by Paul bear the names of the cities or churches he wrote to but also have his statement of authorship. Luke wrote the Gospel bearing his name and Acts and John the Beloved Disciple the Gospel and epistles bearing his name as well as the book of Revelation.

A few books are of uncertain authorship, Esther and Hebrews, for example. Honest disagreement about these matters does not mean the scholars do not believe in the Inspiration of the Scriptures. Textual Critics are another form of legitimate scholars who seek to find the best ancient manuscripts and discover the most accurate meaning of the Scriptures.

By contrast, "Higher Criticism" of the Bible compares ancient documents of many cultures with Scriptures as if the Bible were no different from any other writing. "Higher Criticism treats the Bible as a text created by human beings at a particular historical time and for various human motives, in contrast with the treatment of the Bible as the inerrant word of God," says Wikipedia. Wikipedia's article on the subject contains much confusion between Higher and Textual Criticism and even calls Higher Criticism "Historical" Criticism. For instance, it claims that Desiderus Erasmus (1466?-

1536), publisher of the Greek New Testament Martin Luther used in his translation work, was a Higher Critic. In fact he was a textual critic, simply seeking to find the best biblical manuscripts to obtain the most accurate meanings of the texts. His work was essential to translators like Luther, Calvin and the KJV scholars.

“The School of Higher Criticism” questioned the authorship of most books of the Bible and rejected many as being not part of the Scriptures. They believed all the supernatural events could be explained by natural means and interpreted Scriptures in the light of cultural superstitions of the surrounding nations. They actually held these cultural superstitions of the surrounding nations superior to the Bible because there are archaeological artifacts supporting the fact that these nations believed these superstitions. The lack of artifacts to support Israel’s culture lessens the authenticity of Israel’s culture in the eyes of these “higher critics.” They had no use for any of the Old Testament as spiritual truth but only for what historical information it might contain. Higher Critics included Jean Astruc (mid-18th cent.), a French Physician and the first to begin formulating the JEDP or Documentary Hypothesis (see below). Johann Salomo Semler (1725-91) questioned the equality of authority of the Old and New Testaments, the authorship of most of the books of the Bible, and any true revelation or inspiration by God. Johann Gottfried Eichhorn (1752-1827), a Professor of Oriental Languages, was sometimes called the “founder of modern Old Testament criticism.”

Ferdinand Christian Baur (1792-1860) was a member of the Tübingen School. He argued that "Jewish Christianity" and "Pauline Christianity" were opposing viewpoints in the New Testament and placed the writings of the pastoral epistles much later than conservative scholars.

Julius Wellhausen (1844-1918) was best known for refining the JEDP Documentary Theory. He demanded a much later time period for the writing of the Pentateuch. He dismissed the possibility of early Jewish monotheism and their status as "God's chosen people." Some of his works are believed to have encouraged anti-Semitism and formed the justification for later Nazi beliefs and practices. Friedrich Schleiermacher (1768–1834) made an early confession in a letter to his father about his religious doubts, no doubt speaking for many of the Higher Critics.

"Faith is the regalia of the Godhead, you say. Alas! dearest father, if you believe that without this faith no one can attain to salvation in the next world, nor to tranquility in this - and such, I know, is your belief - oh! then pray to God to grant it to me, for to me it is now lost. I cannot believe that he who called himself the Son of Man was the true, eternal God; I cannot believe that his death was a vicarious atonement."

In his later work, Addresses on Religion (1799), Schleiermacher makes clear his position as one based simply on feelings, with no Scriptural truth or doctrine involved. This is in strange contrast to the fact that he was supposedly a great lecturer and scholar on Hermeneutics, the attempt to examine and arrive at the true meaning of Scriptural texts.

David Friedrich Strauss (1808–74) wrote The Life of Jesus, Historically Examined. It dismissed all possibility that Jesus Christ was the Son of God. It treated the gospels as mythology and claimed they were written by recreating a Jewish Messianic tradition of teaching folklore. Ludwig Feuerbach (1804–72) studied external manuscripts in what he claimed was a quest to confirm events related in the Bible. His works were later translated by George Eliot [see below] in England and taught that man and God are essentially the same, since nature is all that exists and man is a part of nature. Attributes of the Divine are just projections from man himself. He is considered to be a primary inspiration for the Marxist Dialectic. Higher Critics in these later times were inspired by Enlightenment and Rationalist thinkers like John Locke, David Hume, Immanuel Kant, Gotthold Lessing, Gottlieb Fichte, Georg Hegel and the French rationalists. (See the appendix on Natural Law to examine whether some of these writers would have agreed with the position of the Higher Critics on the Scriptures.)

Samuel Taylor Coleridge brought Higher Criticism to England along with George Eliot, who translated Strauss's The Life of Jesus (1846) and Feuerbach's The Essence of Christianity (1854). (Note that George Elliot, or Mary Anne Evans, wrote Silas Marner, a bitter attack on organized religion and staunch advocacy for humanism.) Seven Anglican theologians began in 1860 to make this criticism a part of Christian doctrine in Essays and Reviews. Dealing with this work was considered more important by the Church of that time than the publication of Origin of Species by Darwin and absorbed five years in a fiery battle for the truth.

Two authors of Essays and Reviews were charged with heresy and fired in 1862. By 1864 the judgment was reversed on appeal. La Vie de Jésus (1863), by Ernest Renan (1823–92), followed in the footsteps of Strauss and Feuerbachin, stripping Jesus Christ of His divinity. In Catholicism, Alfred Loisy wrote L'Evangile et l'Eglise (1902), opposing Essence of Christianity of Adolf von Harnack, who questioned early Christian church doctrines, rejected John's Gospel because of the differences with the Synoptic Gospels and promoted the social gospel, saying that man's relationship with God was personal and needed no organized church. Alfred Loisy, however, believed that Jesus Christ did not consciously know He was one with God. La Vie de Jesus by Ernest Renan once again attempted to make Jesus a "real person" while robbing Him of His Divine Nature. Albert Schweitzer criticized the "human only" view of Jesus, but replaced it with the idea that Jesus and his disciples thought an immediate fulfillment of the prophecies of judgment was going to occur, in His lifetime, and only when they didn't occur did He realize He had to die for sin. Schweitzer was still a humanist and apparently reasoned that Jesus' great sacrifice made the fulfillment of the prophecies unnecessary. Rudolf Bultmann, among others, claimed higher criticism was needed to "demythologize" the Bible.

The "Documentary Hypothesis" of Wellhausen, also called the "J, E, D, P Hypothesis," claims to identify supposed writers of the respective portions of the Pentateuch (Genesis-Deuteronomy) especially, and also much of the Old Testament. The "Jehovist Document," supposedly dates from about 850 B.C., used Jehovah for God; the "Elohist Document," about 750 B.C., used Elohim; the "Deuteronomist Document," was supposed to be an editor of the first two, dated about 620 B.C., containing especially most of the Book of Deuteronomy; and, finally, the "Priestly Document," supposed editorial revisions by Jewish priests around 500 B.C.

It was necessary to attribute the Bible to different or unknown authors, especially to place the writings into later time periods, because these "scholars" came to their work presupposing that all ancient manuscripts were equal, that they probably had an even more ancient common source in the "oral tradition," and that the world couldn't be only six to ten thousand years old and have an advanced civilization as described in Genesis four and six when they didn't even credit people from Moses' time with the ability to write. It was also essential to discredit prophecy in the later books of the Old Testament, so their dates of writing were placed after the fulfillment of the prophecy. The Bible can't be reduced to the level of any other book if it accurately predicts the future, so the prophecies can't be allowed to remain in the time period where they were actually made.

Higher Critics can produce a chart showing each Book of the Bible and assigning it an unknown or post-period author/editor. They consign some of the acknowledged epistles to the category of the pseudepigrapha, the non-scriptural writings, like the Apocrypha. The key is the Higher Critics' presuppositions. They were presupposing that there was no God who gave His words to Man. They were presupposing that culture and religion evolved as man evolved, because they were evolutionists, of course, and all cultures and religions were equally mythological. The first eleven chapters of Genesis must have been rehashed and modified from the Babylonian myths, which had to be older than the Bible. The truth is that writing is older than Moses who lived thirty-five hundred years ago, as proved by archaeological finds, and Abraham lived when the Bible says he lived. But the Higher Critics couldn't have Monotheism and metallurgy and city-building and musical instruments where there should be nomadic hunter-gatherers scratching paintings on cave walls.

Christ Himself quoted the Old Testament extensively and credited its acknowledged authors as true prophets: "And beginning at Moses and all the prophets, he expounded unto them in all the scriptures the things concerning himself. . . These are the words which I spake unto you, while I was yet with you, that all things must be fulfilled, which were written in the law of Moses, and in the prophets, and in the psalms, concerning me" (Luke 24:27, 44).

There are still questions about the writings describing creation and events before Moses' birth and after his death. Possibly God taught Moses these directly. It is also possible that manuscripts of earlier writers whom God inspired were collected and preserved, by Noah on the Ark, and others, and handed over to Moses to work from. Possibly Samuel was given an account of Moses' death written by Joshua. The Scriptures do not say, but these theories do not rob the Scriptures of their uniqueness and their truth as Higher Criticism does.

Higher critics, on the other hand, do not believe that any of the books of Old Testament were written in form we have them now before about 250 BC. They believe that the New Testament was not complete until the council of Nicene in 325 AD. One solid example of the stark contrast between conservatives and higher critics is found in the synoptic gospels. Conservatives point out that Church tradition has always held that the Gospel of Matthew was written first, probably from Aramaic sources, maybe as early as 35 AD, certainly no later than 45 AD. Matthew was written to Jewish believers and he expected his readers to have a solid knowledge of the Old Testament. Luke was written soon after Matthew to a Gentile believer who needed some explanations of Jewish traditions. The Gospel of Mark is traditionally known as Peter's Gospel, written when Peter was an old man in Rome, with Mark as transcriber and possibly editor.

The Roman culture did not allow a great deal of leisure time, so Mark's Gospel is cut to about half the size of Matthew or Luke and includes many words to emphasize action, such as immediately and straightway. Higher Critics say that the Gospel of Mark was written first simply because it is the shortest. Without the slightest evidence, higher critics believe that the Gospel of Mark was written from some smaller document which they call "Q." Material was made up and added to make the Gospel of Mark. Still later, more material was added to the Gospel of Mark to create the Gospels of Matthew, Luke and John. According to Higher Critics, "Q" could have been written as late as 45 AD, the Gospel of Mark maybe around 65 AD and the other Gospels as late as the Nicene Council 325

AD. Though this is accepted throughout the academic world and has sold millions of books, this is all without any evidence.

[Some of the material on the Documentary Theory, archaeological proofs and authorships is adapted from The Genesis Record: a Scientific and Devotional Commentary on the Book of Beginnings, by Henry M. Morris, Baker Book House, Grand Rapids, MI.]

Problems with Calendars

Almost everyone on planet Earth works with more than one calendar every single day. The United States of America officially uses the Gregorian calendar. A budget calendar, the one that determines a company or other business entity's working year, is called a Fiscal calendar. The federal budget uses a calendar which begins October 1. The vast majority of school systems or school districts use a calendar which begins around the first of September. Employees of NASA or the US military use an Ordinal calendar which starts approximately the first week of January on a Gregorian calendar and numbers each week without naming months. This type of calendar is very useful for imputing computer data, since the week number determines where you are in the calendar year, not the month. This system has international application as well. Because companies can operate on any calendar they want, the different possible calendar combinations are nearly infinite.

It is quite normal for an American family to use one calendar for the husband's job, a different calendar for the wife's job, different calendars for each school-aged child, a different calendar for taxes, plus an attempt to synchronize these different calendars with a Gregorian calendar. To add to the confusion, it is also normal for every one of these families to know families with completely different calendars. So why do so many people fail to understand the difficulty of synchronizing a modern Gregorian calendar with an ancient calendar?

Modern calendars began with Julius Caesar. When he began collecting taxes throughout his empire, the lack of a common calendar was a serious difficulty. He issued a new calendar to aid in tax collection. Though it was very good, it was not perfect and has been corrected many times throughout the centuries. The standard throughout the Western world today is a Julian calendar revised by Pope Gregory known as a Gregorian calendar. In the middle ages a mathematician developed the Ordinal calendar which is also called a Julian calendar but is not related to Julius Caesar's Julian calendar. A Revised Julian calendar was adopted by the Orthodox Churches of Constantinople, Alexandria, Antioch, Greece, Cyprus, Romania, Poland, and Bulgaria by 1963. This is distinct from previous Julian calendar revisions or the Gregorian calendar. Confused yet?

Realizing how difficult our daily lives are with all of these various calendars, the real question is how do we even come close to matching ancient dates to our Gregorian calendar? Prior to Julius Caesar, each country and often every city had its own calendar. Each new king would introduce his own new calendar. Sometimes two kings would reign at the same time, especially father and son, and two different calendars would overlap. Sometimes the beginning or the end of a king's reign would count a partial year as a full year. Each culture also had civil and religious calendars which started at different times of the year and often had a different number of days in the year. Israel still functions this way today. The book The Mysterious Numbers of the Hebrew Kings by Edwin Thiele does a good job of reconciling the many pieces of information contained in the Bible concerning the dates of the kings from Saul to the Babylonian captivity.

Our position is that modern historians do not do a very good job of synchronizing modern calendars with ancient events before Alexander the Great. Since Alexander, writers throughout the Mediterranean world have used common events to synchronize their calendars, so synchronizing their writings with our calendars is much easier. From the time period of Alexander the Great back to the founding of the Chaldean or Babylonian Empire at the battle of Carchemish under Nabopolasser, Nebuchadnezzer's father, the many common events still allow for a very high degree of synchronization. Before the Battle of Carchemish, however, the calendars of different nations have few common events to synchronize. Since mainstream historians rely on the highly flawed Egyptian timeline popularized by James Breasted in his Ancient Records of Egypt and supplement that timeline with artifacts which are either radiocarbon dated or simply assigned an assumed date, there are plenty of opportunities for massive errors. Immanuel Velikovsky, for instance, in his book Ages In Chaos, believes that synchronization of Egyptian history with Hebrew history at the time of King Saul has Hebrew history too young by about 200 years and

Egyptian history at this point too old by about 300 years for a total error of at least 500 years.

Prejudice against the Bible influences often influences objectivity. To avoid this prejudice, we will examine the founding of Carthage as an example. The founding of Carthage is an example of archaeological dating unrelated to biblical records. Many ancient writings reference the founding of Carthage, such as Virgil's Aeneid and The Antiquities of the Jews by Flavius Josephus. Timaeus of Tauromenion , Dionysius of Halicarnassus and Velleius Paterculus date the founding of Carthage in the same general time period. The number of references to Carthage as a city are staggering, so we can be certain that Carthage was founded sometime. All of the ancient sources date the founding of Carthage in the mid to late ninth century BC (853-814 BC). Carthage was, however, so thoroughly destroyed, first by the Romans in 146 BC and later by the Muslims in 698 AD, that there is almost no physical evidence that Carthage as a Phoenician capital ever existed. The scanty remains which modern archeologists have unearthed are never dated anywhere near the recorded late ninth century date, usually no earlier than 725 BC.

And this is the problem. Modern historians, knowing that the city of Carthage was destroyed many times, place more faith in a few remaining artifacts than the written records. Just because we no longer have any physical evidence that Carthage was founded in the ninth century, modern historians are completely discounting all ancient written records and claiming that Carthage was founded much later. This nonsense has actually become mainstream. Reliance upon physical objects from an archaeological dig as the sole means of dating a site's true age, origin and beginning is neither good science nor good sense. Dating methods can produce flawed results. Artifacts might be from later periods of time than the first settlers. They might be from other cultures, imported in trade or left by invaders.

This is a case where a staggering weight of written evidence is being ignored in favor of a very few artifacts with no certain connection to the founding of Carthage. Scholars around the ancient world acknowledged the founding of Carthage. Archaeologists use the evidence of broken pots or pieces of charcoal instead of written records.

An example of this prevailing prejudice is found in a brief article on the founding of Carthage at varchive.org, a scholarly archive of Immanuel Velikovsky's unpublished works. "The Date of Carthage's Founding," from a supplementary section called, New Light on the Dark Age of Greece by Jan Sammer. Jan Sammer (b. Plzen, Czechoslovakia, 1953) was an assistant to Immanuel Velikovsky (1976-1979), an archivist and editor for the Velikovsky Estate, 1980-1983. He has a Bachelor of Arts from SGWU, Montreal, 1975, and a Master of International Affairs, Columbia University, NYC, 1986. He has also made contributions to Kronos and Aeon journals.

"Archaeology, however, does not support a mid- or late-ninth century date for Carthage's founding. After many years of digging archaeologists have succeeded to penetrate to the most ancient of

Carthage's buildings. P. Cintas, excavating a chapel dedicated to the goddess Tanit, found in the lowest levels a small rectangular structure with a foundation deposit of Greek orientalizing vases datable to the last quarter of the eighth century. These are still the earliest signs of human habitation at the site; although Cintas originally held out hope that there would be found remains of the earliest settlers of the end of the ninth century, the years have not substantiated such expectation. Scholars are now for the most part ready to admit that the ancient chronographers' estimate of the date of the city's founding was exaggerated."

These rejected written records do not even include the Hebrew Scriptures. The bias of archeologists against written records added to the bias against the Bible make a synchronization of calendars impossible.

Even Conservatives who start with the position that the Biblical record is factually accurate have extreme difficulty synchronizing the dates of ancient events with our modern Gregorian calendar. The most important date and the starting point looking backwards from the present is the dating of the Fall of Jerusalem. Since the campaigns of Nebuchadnezzar are recorded in multiple places, the Fall of Jerusalem to the Babylonians in the early sixth century BC is universally accepted. Since the campaign against Jerusalem took several years and we have that campaign mentioned from several points of view, a disagreement over the exact date is almost inevitable. This does not mean, however, that these conservatives disagree about the facts. One conservative might believe the evidence concludes that the final fall of Jerusalem took place in 584 BC of our calendar. Someone else might believe the evidence concludes the final fall of Jerusalem took place in 587 BC. These men have no disagreement over the facts. The only thing they disagree about is method of synchronizing the calendars.

As we go back in time, these minor disagreements over calendar synchronization will grow greater. The correct dating of the Fall of Jerusalem determines the correct dating of the founding of Solomon's temple, which determines the correct dating of the Exodus of Egypt, which determines the correct dating of the call of Abraham, which determines the correct dating of the flood of Noah, which determines the correct dating of the creation of the world.

Specific examples of attacks on recognized members of the scientific community because of a connection with Intelligent Design

Dr. Richard Sternberg, on his website RichardSternberg.org, describes himself as an evolutionary biologist with interests in the relation between genes and morphological homologies, and the nature of genomic 'information.' I hold a Ph.D. in Biology (Molecular Evolution) from Florida International University and a Ph.D. in Systems Science (Theoretical Biology) from Binghamton University.

His position of respect and influence in the mainstream scientific community is undeniable. From 2001-2007, I served as a staff scientist at the National Center for Biotechnology Information, and from 2001-2007 I was a Research Associate at the

Smithsonian's National Museum of Natural History. I am presently a research scientist at the Biologic Institute, supported by a research fellowship from the Center for Science and Culture at Discovery Institute. I am also a Research Collaborator at the National Museum of Natural History.

Dr. Sternberg goes on to say that, In 2004, in my capacity as editor of The Proceedings of the Biological Society of Washington, I authorized "The Origin of Biological Information and the Higher Taxonomic Categories" by Dr. Stephen Meyer to be published in the journal after passing peer-review. Because Dr. Meyer's article presented scientific evidence for intelligent design in biology, I faced retaliation, defamation, harassment, and a hostile work environment at the Smithsonian's National Museum of Natural History that was designed to force me out as a Research Associate there. Federal government employees acting in concert with an outside advocacy group, the National Center for Science Education, took these actions. Efforts were also made to get me fired from my job as a staff scientist at the National Center for Biotechnology Information. Subsequently, there were two federal investigations of my mistreatment, one by the U.S. Office of Special Counsel in 2005, and the other by subcommittee staff of the U.S. House Committee on Government Reform in 2006. Both investigations unearthed clear evidence that my rights had been repeatedly violated.

The author of the article Dr. Sternberg approved for publication in the above reference is Dr. Stephen C. Meyer. His credentials are as follows, taken from the website of the Access Research Network (ARN.org).

Dr. Stephen C. Meyer received his Ph.D. in the History and Philosophy of Science from the University of Cambridge in 1991 for a dissertation on origin-of-life biology and the methodology of the historical sciences. Formerly a geophysicist with the Atlantic Richfield Company, he is currently Director of the Center for Renewal of Science and Culture at Discovery Institute and an Associate Professor of Philosophy at Whitworth College. He is a past recipient of a Rotary International Scholarship, the American Friends of Cambridge scholarship (administered by the Cambridge Commonwealth Trust) and a Templeton Foundation science-religion teaching grant. He has contributed articles to several scholarly books and anthologies including "The History of Science and Religion in the Western Tradition: An Encyclopedia, Darwinism: Science or Philosophy," "Of Pandas and People: The Central Question of Biological Origins," "The Creation Hypothesis: Scientific Evidence for An Intelligent Designer" and "Facets of Faith and Science: Interpreting God's Action in the World." In addition to technical articles on the philosophy of science, he has published many editorial features in newspapers and magazines such as The Wall Street Journal, The Los Angeles Times, The Chicago Tribune and National Review. He has also recently appeared as a guest on several national television programs including PBS's Freedom Speaks and TechnoPolitics, and CNBC's Hardball with Chris Matthews. He is currently working on a book formulating a scientific theory of biological design, which looks specifically at the evidence for design in the encoded information in DNA.

As a proponent of Intelligent Design, Dr. Meyer and the Center for Science and Culture were featured in an article in The American Prospect, Liberal Intelligence, a web magazine describing itself in part as follows:

The American Prospect was founded in 1990 as an authoritative magazine of liberal ideas, committed to a just society, an enriched democracy, and effective liberal politics. ... Our mission, simply put, is to rise to the momentous occasion that confronts all Americans who seek a just society built on our greatest traditions. Contemporary conservatism stands to thwart those traditions; it advances its agenda by way of stealth, fear-mongering, and a massive propaganda apparatus. It is our mission to expose that agenda and the lies that support it. ... We founded the Prospect out of a conviction that the conservative undertow in American political life is profoundly influenced by the dominance of conservative media and think tanks. Our conservative counterparts have played a critical role in pulling the entire national debate to the right. We intend to take it back.

Chris Mooney is, according to the magazine's website, a Prospect senior correspondent and a freelance writer living in Washington, D.C. He focuses on issues at the intersection of science and politics, and has been praised as a "revolutionary mind" by Seed Magazine, which recently commended his "trenchant brand of science-centered commentary." His most recent articles include a Columbia Journalism Review feature story about the problem with "balance" in science coverage and a Boston Globe commentary.

In an article written December 2, 2002, entitled "Survival of the Slickest: How Anti-Evolutionists are Mutating Their Message," Mooney says,

These "Intelligent Design" (ID) theorists, as they call themselves, are epitomized by Stephen C. Meyer, an anti-Darwinian philosopher who made the following appeal to The American Prospect: "People with liberal credentials ought to understand what we're up against. This is an entrenched establishment."

ID theorists posit that living things, due to their organizational complexity and magnificent design, simply must be the creations of some form of intelligence. Where evolutionary biologists see species evolving through a blind process of natural selection acting over millions of years, ID theorists assert that life as we know it simply could not have arisen in such a manner. Furthermore, they claim that this is a scientific observation. ID advocates don't always articulate precisely what sort of intelligence they think should stand in lieu of evolution on textbook pages, but God -- defined in a very nebulous way -- generally outpolls extraterrestrials as the leading candidate.

ID's home base is the Center for Science and Culture at Seattle's conservative Discovery Institute. Meyer directs the center; former Reagan adviser Bruce Chapman heads the larger institute, with input from the Christian supply-sider and former American Spectator owner George Gilder (also a Discovery senior fellow). From this perch, the ID crowd has pushed a "teach the controversy" approach to evolution that closely influenced the Ohio State Board of Education's recently proposed science standards, which would require students to learn how scientists "continue to investigate and critically analyze" aspects of Darwin's theory.

This language may seem innocuous enough, but it clearly allows teachers room to bring up ID if they choose. Moreover, the proposal is insidious because the standards don't ask for the critical analysis of any other bedrock scientific theories, such as plate tectonics or quantum mechanics. Unless there's a shift in the political winds, however, Ohio will finalize the troubling new standards in December.

Mooney does not seem to want anybody to be able to bring up ID, ever. His publication claims that conservative ideals are overcoming liberal ideals. His solution is to stifle the "opposition" by mockery, arousing fear of invasion by religionists posing as scientists, and complaining that they're going to take over if they're not stopped. It is also interesting to compare Dr. Meyer's list of credentials against Christopher Mooney's. Mr. Mooney, however, doesn't list any credentials, just some publication credits. (He is also mentioned below in the *Blogspot* of Professor Jeffrey Shallit, in connection with one of his books and complimented for his researching ability.) Mooney can apparently enjoy the freedom to write and publish his beliefs even without years of study and productivity in recognized scientific circles. This is the freedom enjoyed in America. Yet he would deny the same rights to Dr. Meyer. Note that what he writes are beliefs, as fervently held and staunchly defended as any organized dogma.

Expelled Exposed: Why Expelled Flunks is a website self-described as "created and maintained by the National Center for Science Education." It goes through the movie item by item and tries to disprove all of its claims. The following is an example of the website's treatment of the content of the movie.

Expelled claims that Iowa State University astronomy professor Guillermo Gonzalez was denied tenure because of his views on intelligent design. However, this shows a naïve and distorted understanding of the tenure process at a major research university. The tenure process involves intense scrutiny of a candidate's accomplishments in order to assess his future potential; the beliefs or extra-academic opinions held by the candidate are not a factor.

According to ISU, Gonzalez's tenure decision was based on "refereed publications, his level of success in attracting research funding and grants, the amount of telescope observing time he had been granted, the number of graduate students he had supervised, and most importantly, the overall evidence of future career promise in the field of astronomy." As documented below, the university had grounds to conclude that the early promise of Gonzalez's career was not being met.

In other words, Professor Guillermo Gonzalez failed to measure up to his university's scrutiny of his present work and future prospects. Unfortunately, all of the factors presented as criteria for evaluation are completely controlled by the university and it is circular reasoning to blame failure on the man when the university didn't grant him the means.

"Refereed publications" means the university has to approve what you write. Failing in "attracting funding and grants" means his ideas for research didn't meet with the approval of the powers that control the purse strings. Who allots the amount of telescope time granted? The university. Again, not his choice. Who assigns graduate students? And how can a man have future promise if he is not permitted to pursue the work he deems to have the most future promise? These so-called proofs certainly admit of more than one interpretation. It is easy to stall a man's career when you control his access to publication, equipment, and students and then accuse him of failing to live up to his promise if his promise doesn't fit in with your dogma.

Also, while teaching at ISU, Gonzalez had 17 articles published. Tenure was withheld not because he failed to publish, but because those in authority did not like what he published. In other words, tenure was withheld from Professor Guillermo Gonzalez because of censorship.

Guillermo Gonzalez - publication record (at ISU)

From the Web of Science (Formerly Science Citation Index)

1. Vanture AD, Smith VV, Lutz J, et al.Correlations between lithium and technetium absorption lines in the spectra of galactic S starsPUBLICATIONS OF THE ASTRONOMICAL SOCIETY OF THE PACIFIC 119 (852): 147-155 FEB 2007Times Cited: 02. Gonzalez GThe chemical compositions of stars with planets: A review PUBLICATIONS OF THE ASTRONOMICAL SOCIETY OF THE PACIFIC 118 (849): 1494-1505 NOV 2006Times Cited: 0

3. Tautvaisiene G, Wallerstein G, Geisler D, et al.Chemical abundances in the Sagittarius galaxy: Terzan 7IAU SYMPOSIA 13: 210-210 2005Times Cited: 04. Gonzalez GIndium abundance trends among sun-like starsMONTHLY NOTICES OF THE ROYAL ASTRONOMICAL SOCIETY 371 (2): 781-785 SEP 11 2006Times Cited: 05. Gonzalez GCondensation temperature trends among stars with planetsMONTHLY NOTICES OF THE ROYAL ASTRONOMICAL SOCIETY 367 (1): L37-L41 MAR 21 2006Times Cited: 56. Gonzalez GHabitable zones in the universeORIGINS OF LIFE AND EVOLUTION OF THE BIOSPHERE 35 (6): 555-606 DEC 2005Times Cited: 07. Gonzalez GMisrepresenting intelligent designSCIENTIST 19 (16): 8-8 AUG 29 2005Times Cited: 0

8. Giridhar S, Lambert DL, Reddy BE, et al.

Abundance analyses of field RV Tauri stars. VI. An extended sampleASTROPHYSICAL JOURNAL 627 (1): 432-445 Part 1 JUL 1 2005Times Cited: 89. Geisler D, Smith VV, Wallerstein G, et al. “Sculptoring” the galaxy? The chemical compositions of red giants in the Sculptor dwarf spheroidal galaxyASTRONOMICAL JOURNAL 129 (3): 1428-1442 MAR 2005Times Cited: 1610. Laws C, Gonzalez GA reevaluation of the super-lithium-rich star in NGC 6633ASTROPHYSICAL JOURNAL 595 (2): 1148-1153 Part 1 OCT 1 2003Times Cited: 511. Laws C, Gonzalez G, Walker KM, et al.Parent stars of extrasolar planets. VII. New abundance analyses of 30 systemsASTRONOMICAL JOURNAL 125 (5): 2664-2677 MAY 2003Times Cited: 4812. Wells LE, Armstrong JC, Gonzalez GReseeding of early Earth by impacts of returning ejecta during the late heavy bombardmentICARUS 162 (1): 38-46 MAR 2003Times Cited: 1313. Gonzalez GColloquium: Stars, planets, and metalsREVIEWS OF MODERN PHYSICS 75 (1): 101-120 JAN 2003Times Cited: 2714. Candia P, Krisciunas K, Suntzeff NB, et al.Optical and infrared photometry of the unusual Type Ia supernova 2000cxPUBLICATIONS OF THE ASTRONOMICAL SOCIETY OF THE PACIFIC 115 (805): 277-294 MAR 2003Times Cited: 2315. Armstrong JC, Wells LE, Gonzalez GRummaging through Earth's attic for remains of ancient life ICARUS 160 (1): 183-196 NOV 2002Times Cited: 1216. Reddy BE, Lambert DL, Laws C, et al.A search for Li-6 in stars with planetsMONTHLY NOTICES OF THE ROYAL ASTRONOMICAL SOCIETY 335 (4): 1005-1016 OCT 1 2002Times Cited: 2617. Gonzalez G, Brownlee D, Ward PDRefugees for life in a hostile universeSCIENTIFIC AMERICANTimes Cited: 1

Pamela Winnick, a former writer for the Pittsburgh Post-Gazette, is interviewed in Expelled because she wrote a piece mentioning Intelligent Design and claimed that she was blacklisted as a journalist because of it. Jeffrey Shallitt, a Canadian professor, gives space in his "Blogspot" to discussing Winnick.

Monday, July 10, 2006 "Pamela Winnick's Science Envy"

Pamela Winnick is an attorney and former reporter for the Pittsburgh Post-Gazette who has written several articles that lean against evolution and in favor of intelligent design. I recently forced myself to read her 2005 book, A Jealous God: Science's Crusade Against Religion. It wasn't a pleasant experience. Winnick's book covers a variety of topics: abortion, population control, eugenics, medical experimentation, the Scopes trial, the theory of evolution, intelligent design, and fetal tissue research. Her thesis -- if this rambling, disjointed book can be said to have one -- is contained in the book's final paragraph ... "The Galileo prototype of the scientist martyred by religion is now purely a myth. Science long ago won its war against religion, not just traditional religion, but any faith in a power outside the human mind. Now it wants more.

"Throughout the book, scientists are depicted as crazed, power-hungry, and immoral. Only religion, Winnick implies, can rein in these dangerous nuts who threaten society. Winnick's claim that "science long ago won its war against religion" is far too glib. Ironically, 2005 also saw the publication of Chris Mooney's The Republican War on Science, a far-better-documented book that shows in depressing detail how American science has been subjugated to the political and especially religious goals of the Christian right. Winnick's reporting is sloppy. Incidents are slanted to support her thesis, names are misspelled (Stanislaw Ulam's last name is comically morphed into "Ulsam"; Richard Lewontin's middle initial is given incorrectly), quotes are mined (sometimes incorrectly), and some "facts" are just plain made up.

Detractors claim that Winnick hasn't been blacklisted. She has continued to write for pay and was in no way harmed by those who object to her giving space to ID.

Certainly she has written for publication since this blacklisting. But consider that a person in an occupation needs to make a full-time living at that occupation. He or she must be able to sell enough writing to live from the proceeds. Listing a handful of works she has sold, including a book, does not prove that she has not been censured. It seems to prove the opposite.

If this were a Marxist, totalitarian regime she would of course have published nothing. As it is, she has probably managed to find a smaller voice and no doubt a smaller living. The attacks in the foregoing quote represent one voice among many who have risen to censure her for daring to give a voice to ID thinking. Note the condescending tone, the insults, the insertion of a plug for the "right" (or is it "left"?) kind of writing. He accuses Winnick of being sloppy, slanted, picking at minor errors, which may not even have been her doing, but editorial responsibilities, "mining quotes," as if a person can afford to include every nuance of every scholar's perspective on a topic between the covers of a book that isn't his work, and even calls her a liar. He provides documentation for many of his claims, of course, but as was pointed out in the Professor Guillermo Gonzalez situation, there are multiple perspectives to consider when presenting evidence, and if anybody's perspective is slanted, it is Professor Jeffrey Shallit's.

Discovery Institute Background, Positions and Membership

(Author's Note: All of the material following comes directly from the Discovery Institute website, www.discovery.org, relating to the PBS series Evolution which aired in 2001. This material is included in its entirety to explain what the

Discovery Institute is, to make clear who its members and associates are, and what they do and do not believe, in their own words. Special notice should be taken of the Zogby Poll released at the same time as this report and its conclusions about what Americans think about the teaching of Evolution or the introduction of Intelligent Design evidence in the classroom. The complete poll is available for review on Discovery Institute's website.) Discovery Institute is a non-profit, non-partisan public policy center for national and international affairs. The Institute's mission is to make a positive vision of the future practical. The Institute discovers and promotes ideas in the common sense tradition of representative government, the free market and individual liberty. Its mission is promoted through books, reports, legislative testimony, articles, public conferences and debates, plus media coverage and the Institute's own publications and website. Current projects explore the fields of technology, science and culture, reform of the law, national defense, the environment and the economy, the future of democratic institutions, transportation, religion and public life, government entitlement spending, foreign affairs and cooperation within the bi-national region of "Cascadia". Discovery Institute 1511 Third Ave Suite 808 Seattle, WA 98101 (206) 292-0401 fax (206) 682-5320 *http://www.discovery.org*

Discovery Institute's Critique of the PBS 2001 series Evolution:

SEATTLE--In an ironic greeting to the seven-part public television series "Evolution" that begins tonight, 100 scientists have declared that they "are skeptical of claims for the ability of random mutation and natural selection to account for the complexity of life." The signers say, "Careful examination of the evidence for Darwinian theory should be encouraged."

Discovery Institute, a Seattle-based public policy center, compiled the list of statement signers (attached). Among other things, the long list may help to answer the contention of designated spokespeople for the series "Evolution" that "virtually all reputable scientists in the world" support Darwin's theory. Institute officials charge that officials of WGBH/Clear Blue Sky Productions have used that contention to keep any scientific criticism of Darwinism from being acknowledged or examined in the eight-hour series. "They want people to think that the only criticism of Darwin's theory today is from religious fundamentalists," said Discovery president Bruce Chapman. "They routinely try to stigmatize scientists who question Darwin as 'creationists'. "Chemist and five time Nobel nominee, Henry "Fritz" Schaefer of the University of Georgia, commented on the need to encourage debate on Darwin's theory of evolution. "Some defenders of Darwinism," says Schaefer, "embrace standards of evidence for evolution that as scientists they would never accept in other circumstances." Schaefer was on the roster of signers of the statement, termed "A Scientific Dissent on Darwinism." Meanwhile, a Zogby Poll released today shows overwhelming public support--81 percent--for the position that "When public broadcasting networks discuss Darwin's theory of evolution, they should present the scientific evidence for it, but also the scientific evidence against it." Only 10 percent support presenting "only the scientific evidence that supports" Darwin's theory. (Less than 10 percent said "Neither" or "Not sure.") "Public television producers are clearly at odds with overwhelming public sentiment in favor of hearing all scientific sides of the debate," said Chapman, a former Director of the US Census Bureau. "The huge majorities in the poll cross every demographic, regional and political line in America." The national sample of 1,202 adults was conducted by Zogby International from August 25-29. The margin of error is +/-3.0%. Discovery Institute commissioned the Zogby poll, though the survey itself was designed by the Zogby organization. It also included questions on education and "intelligent design," a theory that some scientific critics of Darwin support. (That theory makes no religious claims, but says that the best natural evidence for life's origins points to design rather than a process of random mutation and natural selection.) Discovery Institute last week also opened a special website (www.reviewevolution.org) to critique the WGBH/Clear Blue Sky series in a scholarly "Viewer's Guide." Discovery officials say that the website analyzes all program segments in the series and has uncovered numerous scientific and historical errors, exaggerations and omissions. Full results of the Zogby poll also are available on the website. "The numbers of scientists who question Darwinism is a minority, but it is growing fast," said Stephen Meyer, a Cambridge-educated philosopher of science who directs the Center for the Renewal of Science and Culture at Discovery Institute. "This is happening in the face of fierce attempts to intimidate and suppress legitimate dissent. Young scientists are threatened with deprivation of tenure. Others have seen a consistent pattern of answering scientific arguments with ad hominem attacks. In particular, the series' attempt to stigmatize all critics--including scientists--as religious 'creationists' is an excellent example of viewpoint discrimination." Signers of the statement questioning Darwinism came from throughout the US and from several other countries, representing biology, physics, chemistry, mathematics, geology, anthropology and other scientific fields. Professors and researchers at such universities as Princeton, MIT, U Penn, and Yale, as well as smaller colleges and the National Laboratories at Livermore, CA and Los Alamos, N.M., are included. A number of the signers have authored or contributed to books on issues related to evolution, or have books underway. Despite repeated requests, the series' producers refused to cover scientific objections to Darwinism. Instead, the producers offered only to let scientific dissenters go on camera to tell their "personal faith stories" in the last program of the series, "What About God?" According to Discovery's Chapman, "This was almost an insult to serious scientists. Some of these dissenting scientists are not even religious. When you watch that last program, you realize they were wise to refuse to take part in it." Jed Macosko, a young research molecular biologist at the University of California, Berkeley, and a statement signer, said, "It is time for defenders of Darwin to engage in serious dialogue and debate with their scientific critics. Science can't grow where institutional gatekeepers try to prevent new challengers from being heard."

A Scientific Dissent on Darwinism "I am skeptical of claims for the ability of random mutation and natural selection to account for the complexity of life. Careful examination of the evidence for Darwinian theory should be encouraged." Henry F.Schaefer: Director, Center for Computational Quantum Chemistry: U. of Georgia • Fred Sigworth: Prof. of Cellular & Molecular Physiology- Grad. School: Yale U. • Philip S. Skell: Emeritus Prof. Of Chemistry: NAS member • Frank Tipler: Prof. of Mathematical Physics: Tulane U. • Robert Kaita: Plasma Physics Lab: Princeton U. • Michael Behe: Prof. of Biological Science: Lehigh U. • Walter Hearn: PhD Biochemistry-U of Illinois • Tony Mega: Assoc. Prof. of Chemistry: Whitworth College • Dean Kenyon: Prof. Emeritus of Biology: San Francisco State U. • Marko Horb: Researcher, Dept. of Biology & Biochemistry: U. of Bath, UK • Daniel Kubler: Asst. Prof. of Biology: Franciscan U. of Steubenville • David Keller: Assoc. Prof. of Chemistry: U. of New Mexico • James Keesling: Prof. of Mathematics: U. of Florida • Roland F. Hirsch: PhD Analytical Chemistry-U. of Michigan • Robert Newman: PhD Astrophysics-Cornell U. • Carl Koval: Prof., Chemistry & Biochemistry: U. of Colorado, Boulder • Tony Jelsma: Prof. of Biology: Dordt College • William A.Dembski: PhD Mathematics-U. of Chicago: • George Lebo: Assoc. Prof. of Astronomy: U. of Florida • Timothy G. Standish: PhD Environmental Biology-George Mason U. • James Keener: Prof. of Mathematics & Adjunct of Bioengineering: U. of Utah • Robert J. Marks: Prof. of Signal & Image Processing: U. of

Washington • Carl Poppe: Senior Fellow: Lawrence Livermore Laboratories • Siegfried Scherer: Prof. of Microbial Ecology: Technische Universitaet Muenchen • Gregory Shearer: Internal Medicine, Research: U. of California, Davis • Joseph Atkinson: PhD Organic Chemistry-M.I.T.: American Chemical Society, member • Lawrence H. Johnston: Emeritus Prof. of Physics: U. of Idaho • Scott Minnich: Prof., Dept of Microbiology, Molecular Biology & Biochem: U. of Idaho • David A. DeWitt: PhD Neuroscience-Case Western U. • Theodor Liss: PhD Chemistry-M.I.T. • Braxton Alfred: Emeritus Prof. of Anthropology: U. of British Columbia • Walter Bradley: Prof. Emeritus of Mechanical Engineering: Texas A & M •

Paul D. Brown: Asst. Prof. of Environmental Studies: Trinity Western U. (Canada) • Marvin Fritzler: Prof. of Biochemistry & Molecular Biology: U. of Calgary, Medical School • Theodore Saito: Project Manager: Lawrence Livermore Laboratories • Muzaffar Iqbal: PhD Chemistry-U. of Saskatchewan: Center for Theology the Natural Sciences • William S. Pelletier: Emeritus Distinguished Prof. of Chemistry: U. of Georgia, Athens • Keith Delaplane: Prof. of Entomology: U. of Georgia • Ken Smith: Prof. of Mathematics: Central Michigan U. • Clarence Fouche: Prof. of Biology: Virginia Intermont College • Thomas Milner: Asst. Prof. of Biomedical Engineering: U. of Texas, Austin • Brian J.Miller: PhD Physics-Duke U. • Paul Nesselroade: Assoc. Prof. of Psychology: Simpson College • Donald F.Calbreath: Prof. of Chemistry: Whitworth College • William P. Purcell: PhD Physical Chemistry-Princeton U. • Wesley Allen: Prof. of Computational Quantum Chemistry:

U. of Georgia • Jeanne Drisko: Asst. Prof., Kansas Medical Center: U. of Kansas, School of Medicine • Chris Grace: Assoc. Prof. of Psychology: Biola U. • Wolfgang Smith: Prof. Emeritus-Mathematics: Oregon State U. • Rosalind Picard: Assoc. Prof. Computer Science: M.I.T. • Garrick Little: Senior Scientist, Li-Cor: Li-Cor • John L. Omdahl: Prof. of Biochemistry & Molecular Biology: U. of New Mexico • Martin Poenie: Assoc. Prof. of Molecular Cell & Developmental Bio: U. of Texas, Austin • Russell W.Carlson: Prof. of Biochemistry & Molecular Biology: U. of Georgia • Hugh Nutley: Prof. Emeritus of Physics & Engineering: Seattle Pacific U. • David Berlinski: PhD Philosophy-Princeton: Mathematician, Author • Neil Broom: Assoc. Prof., Chemical & Materials Engineeering: U. of Auckland • John Bloom: Assoc. Prof., Physics: Biola U. • James Graham: Professional Geologist, Sr. Program Manager: National Environmental Consulting Firm • John Baumgardner: Technical Staff, Theoretical Division: Los Alamos National Laboratory • Fred Skiff: Prof. of Physics: U. of Iowa • Paul Kuld: Assoc. Prof., Biological Science: Biola U. • Yongsoon Park: Senior Research Scientist: St. Luke's Hospital, Kansas City • Moorad Alexanian: Prof. of Physics: U. of North Carolina, Wilmington • Donald Ewert: Director of Research Administration: Wistar Institute • Joseph W. Francis: Assoc. Prof. of Biology: Cedarville U. • Thomas Saleska: Prof. of Biology: Concordia U. • Ralph W. Seelke: Prof. & Chair of Dept. of Biology & Earth Sciences: U. of Wisconsin, Superior • James G. Harman: Assoc. Chair, Dept. of Chemistry & Biochemistry: Texas Tech U. • Lennart Moller: Prof. of Environmental

Medicine, Karolinska Institute: U. of Stockholm • Raymond G. Bohlin: PhD

Molecular & Cell Biology-U. of Texas: • Fazale R. Rana: PhD Chemistry-Ohio U. • Michael Atchison: Prof. of Biochemistry: U. of Pennsylvania, Vet School • William S. Harris: Prof. of Basic Medical Sciences: U. of Missouri, Kansas City • Rebecca W. Keller: Research Prof., Dept. of Chemistry: U. of New Mexico • Terry Morrison: PhD Chemistry-Syracuse U. • Robert F. DeHaan: PhD Human Development-U. of Chicago • Matti Lesola: Prof., Laboratory of Bioprocess Engineering: Helsinki U. of Technology • Bruce Evans: Assoc. Prof. of Biology: Huntington College • Jim Gibson: PhD Biology-Loma Linda U. • David Ness: PhD Anthropology-Temple U. • Bijan Nemati: Senior Engineer: Jet Propulsion Lab (NASA) • Edward T. Peltzer: Senior Research Specialist: Monterey Bay Research Institute • Stan E. Lennard: Clinical Assoc. Prof. of Surgery: U. of Washington • Rafe Payne: Prof. & Chair, Biola Dept. of Biological Sciences: Biola U. • Phillip Savage: Prof. of Chemical Engineering: U. of Michigan • Pattle Pun: Prof. of Biology: Wheaton College • Jed Macosko: Postdoctoral Researcher-Molecular Biology: U. of California, Berkeley • Daniel Dix: Assoc. Prof. of Mathematics: U. of South Carolina • Ed Karlow: Chair, Dept. of Physics: LaSierra U. • James Harbrecht: Clinical Assoc. Prof.: U. of Kansas Medical Center • Robert W. Smith: Prof. of Chemistry: U. of Nebraska, Omaha • Robert DiSilvestro: PhD Biochemistry-Texas A & M U., Professor, Human Nutrition, Ohio State University • David Prentice: Prof., Dept. of Life Sciences: Indiana State U. • Walt Stangl: Assoc. Prof. of

Mathematics: Biola U. • Jonathan Wells: PhD Molecular & Cell Biology-U. of California, Berkeley: • James Tour: Chao Prof. of Chemistry: Rice U. • Todd Watson: Asst. Prof. of Urban & Community Forestry: Texas A & M U. • Robert Waltzer: Assoc. Prof. of Biology: Belhaven College • Vincente Villa: Prof. of Biology: Southwestern U. • Richard Sternberg: Pstdoctoral Fellow, Invertebrate Biology: Smithsonian Institute • James Tumlin: Assoc. Prof. of Medicine: Emory U. Charles Thaxton: PhD Physical Chemistry-Iowa State U.

Lunar Recession

PUBLIC INFORMATION OFFICE

JET PROPULSION LABORATORY

CALIFORNIA INSTITUTE OF TECHNOLOGY

NATIONAL AERONAUTICS AND SPACE ADMINISTRATION

PASADENA, CALIF. 911909. TELEPHONE (818) 354-5011

Contact: Diane Ainsworth

FOR IMMEDIATE RELEASE July 21, 1994

During their brief moon walk 25 years ago, the Apollo 11 astronauts deployed a variety of scientific experiments, including a reflector array left in the fine powder of the Sea of Tranquility that continues to measure the moon's orbit around Earth to unprecedented accuracy.

Scientist who analyze data from the Lunar Laser Ranging Experiment have reported some watershed results from these long-term experiments, said Jet Propulsion Laboratory team investigator Dr. Jean

Dickey. The team's findings appear in this week's issue of Science magazine, which commemorates the silver anniversary of the Apollo 11 lunar landing.

"Using the Lunar Laser Ranging Experiment, we have been able to improve, by orders of magnitude, measurements of the moon's rotation," Dickey said. "We also have strong evidence that the moon has a liquid core, and laser ranging has allowed us to determine with great accuracy the rate at which the moon is gradually receding from the Earth."

The laser ranging retroreflector was positioned on the moon in 1969 by the Apollo 11 astronauts so that it would point toward Earth and be able to reflect pulses of laser light fired from the ground. By beaming laser pulses at the reflector, scientists have been able to determine the round-trip travel time of a laser pulse and provide the distance between these two bodies at any given time down to an accuracy of about 3 centimeters (about 1 inch).

The laser reflector consists of 100 fused silica half-cubes, called corner cubes, mounted in a 46-centimeter (18-inch) square aluminum panel.

Each corner cube is 3.8 centimeters (1.5 inches) in diameter. Corner cubes reflect a beam of light directly back toward the point of origin and, thus, allow scientists to measure the Earth-moon separation and study the dynamics of the Earth, the moon and the Earth-moon system.

Once the laser ranging experiments bean to yield valuable results, more reflectors were left on the moon. A reflector identical to the Apollo 11 mission reflector was left the Apollo 14 crew, and a larger reflector using 300 corner cubes was placed on the moon by the Apollo 15 astronauts. French-built reflectors were also left on the moon by the unmanned Russian Lunakhod 2 mission.

Several observatories have regularly ranged the moon with these reflectors: one is located at McDonald Observatory near Fort Davis, Texas; another is located atop the extinct Haleakala volcano on the island of Maui in Hawaii; another is located in southern France near Grasse.

The Lick Observatory in northern California also has been used in the past for the lunar laser ranging experiments and ranging programs have been carried out in Australia, Russia and Germany. Despite the difficulty of detecting reflected laser light from the moon, Dickey said, more than 8,300 ranges have been measured over the last 25 years.

Lunar ranging involves sending a laser beam through an optical telescope," Dickey said. "The beam enters the telescope where the eye piece would be, and the transmitted beam is expanded to become the diameter of the main mirror, then bounced off the surface toward the reflector on the moon."

The reflectors are too small to be seen from Earth, so even when the beam is precisely aligned in the telescope, actually hitting a lunar retroreflector array is technically challenging. At the moon's surface the beam is roughly four miles wide. Scientists liken the task of aiming the beam to using a rifle to hit a moving dime two miles away.

Once the laser beam hits a reflector, scientists at the ranging observatories use extremely sensitive filtering and amplification equipment to detect the return signal, which is far too weak to be seen with the human eye. Even under good atmospheric viewing condition, only one photon-the fundamental particle of light-will be received every few seconds.

The range accuracy of these reflectors has been improved over the lifetime of the lunar laser ranging experiments, the team noted in Science. While the earliest ranges had accuracies of several meters (or several yards), continuing improvements in the lasers and the detection electronics have led to recent measurements that are accurate to about 3 centimeters (about 1 inch).

From the ranging experiments, scientists know that the average distance between the centers of the Earth and moon is 385,000 kilometers (239,000 miles), showing that modern lunar ranges have relative accuracies of better than one part in 10 billion.

This level of accuracy represents one of the most precise distance measurements ever made," Dickey said. "The degree of accuracy is equivalent to determining the distance between Los Angeles and New York to one fiftieth of an inch."

Laser ranging has also made possible a wealth of new information about the dynamics and structure of the moon. Among many new observations, scientists now believe that the moon may harbor a liquid core. The theory has been proposed from data on the moon's rate of rotation and very slight bobbing motions caused by gravitational forces from the sun and Earth.

Other recent findings from the laser ranging experiments include:

--Verification of Einstein's theory of relativity, which states that all bodies fall with the same acceleration regardless of their mass.

--The length of an Earth day has distinct small-scale variations, changing by about one thousandth of a second over the course of a year. These changes are caused by the atmosphere, tides and the Earth's core.

--Precise positions of the laser ranging observatories on Earth are slowly drifting as the crustal plates on Earth drift. The observatory on Maui is seen to be drifting away from the observatory in Texas.

--Ocean tides on Earth have a direct influence on the moon's orbit.

Measurements show that the moon is receding from Earth at a rate of about 3.8 centimeters (1.5 inches) per year.

--Lunar ranging has greatly improved scientists' knowledge of the moon's orbit, enough to permit accurate analyses of solar eclipses as far back as 1400 BC.

Natural Law and its Influence on Science, History and Belief

Aristotle's thinking on the subject is confusing, and he has been called the Father of Natural Law. He seems in his work Rhetoric to have distinguished between particular laws that each nation has set up for itself and common law, that is according to nature. It may be that Aristotle only thought that such a principle could be invoked if the particular laws were in opposition to what the person was trying to obtain in the way of justice.

Aristotle's distinction seemed to indicate a "higher" law, a law of nature. Some people acknowledged a concept called divine positive legislation, and said that laws emanated from the gods. The Stoics believed the universe had purpose and order apart from divine or natural influences, an "eternal" law. To them the true natural law was a rational being living according to this law. Actions had to follow dictates of virtue. Stoics believed that the worth of the individual, his moral duty, and the brotherhood of men were universal ideas. Roman jurists followed these principles, and they were passed down through the development of legal theory.

Many churchmen attempted to incorporate this supposedly secular theory into Christian doctrine. It was difficult to do so without creating confusion about clear teaching on man's fallen nature after the original sin and his inability as a consequence to make or follow any natural laws. Thomas Aquinas believed that man's reason could only go so far and needed God's law to complete it.

Aquinas' beliefs became part of Roman Catholic teaching, the basis for the ideas of the unity of the body and the soul, the conscience, and the ability to perceive and choose good or evil were codified by his teachings. Aquinas believed reason could bring man closer to conformity with God's will. A sort of divine corollary to the idea of preserving life is his belief that reason will lead man to seek good and avoid evil. He believed sin could blind reason and render it ineffective.

There belongs to the natural law, first, certain most general precepts, that are known to all; and secondly, certain secondary and more detailed precepts, which are, as it were, conclusions following closely from first principles. As to those general principles, the natural law, in the abstract, can nowise be blotted out from men's hearts. But it is blotted out in the case of a particular action, insofar as reason is hindered from applying the general principle to a particular point of practice, on account of concupiscence or some other passion, as stated above (77, 2). But as to the other, i.e., the secondary precepts, the natural law can be blotted out from the human heart, either by evil persuasions, just as in speculative matters errors occur in respect of necessary conclusions; or by vicious customs and corrupt habits, as among some men, theft, and even unnatural vices, as the Apostle states (Rm. i), were not esteemed sinful. (Summa Theologica Sixth Article [I-II, Q. 94, Art. 6] Objection 3)

Application has been made to every kind of behavior, and how it relates to man's ability to live by natural law. Even motive must play a part, since doing a good deed for the wrong reason negates the good deed. Aquinas believed that even good intentions were not enough, and he formulated the Cardinal Virtues (Prudence, Justice, Temperance and Fortitude, developed by reason applied to nature). He also believed in the necessity of practicing the theological virtues (Faith, Hope and Charity, developed by spiritual application). All are necessary.

Thomas Hobbes created a massive body of works including one called Leviathan, and in his writings outlined principles that resulted in a natural law based on rational needs for survival, prosperity, and the actions that would need to result from these considerations. Hobbes believed man needed a supreme ruler as the source of law in order for natural law to work. Hobbes essentially believed that natural law forbade man to engage in self-destructive behavior. Man could not take away what preserved life and he had to give up anything that might be harmful to this key principle, that life must be preserved.

Hobbes was a philosopher, and his work led to the strengthening of the power of absolute rulers. He didn't consider whether a ruler was moral or not to be important. Divine Right, the idea that a monarch's power to rule is given by God and cannot be challenged, extends from this philosophy.

Hugo Grotius believed even an all-powerful God could not change natural law. He said that even if there were no God, or if God did not care about man (neither of which he believed was true) natural law would remain solid and unchangeable. This statement gave rise to the separation of natural law from theology. Whether Grotius intended this consequence is questionable. He believed it was impossible that God did not exist, according to his De iure belli ac pacis, Prolegomeni XI), which deals with the natural law question. It seems he only intended to say that natural law was a protective device that would not change and could be relied on, not that since it existed there was no need to consider God.

John Locke took the opposite view from Hobbes and the Divine Right theories. In works that included Two Treatises of Government, he taught that a ruler could violate natural law and his mandate to protect "life, liberty, and property." In such a case the people were obligated to depose and replace him.

Thomas Jefferson's concept of inalienable rights comes from Locke. In the declaration of Independence he refines the natural rights theory and makes it a part of American History. We hold these truths to be self-evident, that all men are created equal, that they are endowed by their Creator with certain unalienable Rights, that among these are Life, Liberty and the pursuit of Happiness.

Pierre Charron in his De la sagesse (1601), said: "The sign of a natural law must be the universal respect in which it is held, for if there was anything that nature had truly commanded us to do, we would undoubtedly obey it universally: not only would every nation respect it, but every individual. Instead there is nothing in the world that is not subject to contradiction and dispute, nothing that is not rejected, not just by one nation, but by many; equally, there is nothing that is strange and (in the opinion of many) unnatural that is not approved in many countries, and authorized by their customs."

This is an interesting contrast to all those who assume the existence of natural law. It is surprising that more secularists and even believers do not hold this view. Charron seems to put his finger on the problem of a fallen world and fallen man exactly. Whether he is himself a believer in fallen nature, he states the truth that we cannot practice obedience to any natural law because our sinful minds are in chaos and conflict.

U.S. statesman James Madison believed that some rights, such as trial by jury, are social rights. These simply develop as social interaction gives rise to a government system and develops laws it needs to exist as a unit. This idea combines the ideas of natural and concrete man-made laws and gives a key to understanding the original intention of the founders in the so-called controversy over the relationship they intended between religion and government.

The following includes excerpts from Dr. Jonathan Sarfati's article "Who's Really Pushing Bad Science?" From Creation.com, the website of Creation Ministries International, including the website's statement of the author's credentials. The quoted material is extensive because it so clearly and completely defines important aspects of natural law, materialism, Humanism, evolution and creation study. It also addresses the issues brought up by those who insist Science must rely on Natural Law and that belief in God has no part in scientific study.

The first paragraph is the website's presentation of Dr. Sarfati's credentials. Dr. Sarfati's article appears in italics, as does all quoted material. He also quotes extensively from the work Good Science, Bad Science: Teaching Evolution in the States by Lawrence S. Lerner, since Sarfati's article is written in response to that work. Both Sarfati and Lerner have outside sources quoted within their works.

Dr. Sarfati studied science at Victoria University of Wellington. He obtained a B.Sc. (Hons.) in

Chemistry with two physics papers substituted (nuclear and condensed matter physics). His Ph.D. in Chemistry was awarded for a thesis entitled 'A Spectroscopic Study of some Chalcogenide Ring and Cage Molecules'. He has co-authored papers in mainstream scientific journals on high temperature superconductors and selenium-containing ring and cage-shaped molecules. He also had a co-authored paper on high-temperature superconductors published in Nature when he was 22.

The following information on Dr. Lerner is taken from the website of the California State University, Long Beach.

Dr. Lawrence S. Lerner is Professor Emeritus of Physics and Astronomy at the California State University, Long Beach. He received his B.A., M.S. and Ph.D. from the University of Chicago. His areas of Expertise are Condensed-matter physics; history of science; science education (college level and K-12); textbooks; science and the humanities; science and religion; K-12 science textbooks, evaluation of K-12 science. His books include two college-level calculus-based physics texts; translation of Giordano Bruno, The Ash Wednesday Supper (1584), An Appraisal of Science Standards in 36 States (1998), and has written more than 100 papers on the subjects listed above. He is also a consultant to several states on science standards.

A Response to Good Science, Bad Science: Teaching Evolution in the States by Lawrence S. Lerner, Thomas B. Fordham Foundation, 26 September 2000.

The theory that Lerner and other materialists are really promoting, and which creationists oppose, is the idea that particles turned into people over time, without any need for an intelligent designer. This 'General Theory of Evolution' (GTE) was defined by the evolutionist Kerkut as 'the theory that all the living forms in the world have arisen from a single source which itself came from an inorganic form.'

Lerner claims that evolution occupies a 'central place' and has a 'unifying role' in the life sciences, and the title even hints that the physical sciences are affected.

Author's Note: Dr. Sarfati refers to the two following Scripture passages in his article. They are placed here in their entirety for easy reference.

(Colossians 1:15–17: He is the image of the invisible God, the firstborn over all creation. For by him all things were created: things in heaven and on earth, visible and invisible, whether thrones or powers or rulers or authorities; all things were created by him and for him. He is before all things, and in him all things hold together.)

(Genesis 2:1–3: Thus the heavens and the earth were completed in all their vast array. By the seventh day God had finished the work he had been doing; so on the seventh day he rested from all his work. And God blessed the seventh day and made it holy, because on it he rested from all the work of creating that he had done.)

Kansas State University immunologist Scott Todd asserted:

'Even if all the data point to an intelligent designer, such an hypothesis is excluded from science because it is not naturalistic.'

That is, never mind the facts — nature is all there is. Naturalism is king! So the opposition to creation has nothing to do with the facts, but with the fact that creationists refuse to play by the self-serving rules of the game formulated by materialists. This contrasts with what most people might think, e.g. double Noble laureate Linus Pauling: 'Science is the search for the truth.'

So how do Lerner et al. try to get around the charge that evolution is really pushing the religion of humanism? After all, the first two tenets of the Humanist Manifesto II (1973), signed by many prominent evolutionists, are:

Religious humanists regard the universe as self-existing and not created.

Humanism believes that Man is a part of nature and has emerged as a result of a continuous process.

The current version, Humanist Manifesto 2000, was signed by the prominent evolutionary propagandists Richard Dawkins, E.O. Wilson, Richard Leakey, Molleen Matsumara and Daniel Dennett.

The atheistic Marxist evolutionist Stephen Jay Gould has claimed that religion and science are 'non-overlapping magisteria' (NOMA). That is, science deals with facts of the real world, while religion deals with ethics, values, morals, and what it means to be human. He expounded this thesis in his book Rocks of Ages: Science and Religion in the Fullness of Life (Ballantyne, NY, 1999).

However, this is based on the philosophically fallacious fact-value distinction, and is really an anti-Christian claim. For example, the Resurrection of Christ is an essential part of the Christian faith (1 Corinthians 15:12–19), but it is also a matter of history, it passed the 'testable' claim that the tomb would be empty on the third day, and impinges on science because it demonstrated the power of God over so-called 'natural laws' that dead bodies decay.

Lerner and others claim scientists must practise methodological naturalism, i.e. that natural causes are the only ones allowed, and God, if He exists, did nothing that can be investigated (contrary to Romans 1:18–23). They claim that doesn't necessarily imply ontological naturalism, i.e. that nature is all that really does exist, and God doesn't. The scare tactic they use to promote methodological naturalism is reasoning like:

'It is simply not possible to solve a scientific question if one is willing to invoke a supernatural answer, because supernatural answers foreclose further scientific inquiry. As we have already noted, a person who accounts for the motion of the planets by asserting that angels propel them is simply not going to be able to account for Kepler's laws of planetary motion in any kind of fruitful way.'

This fails to note the distinction between normal (operational) science, and origins or historical science. Normal (operational) science deals only with repeatable observable processes in the present, while origins science helps us to make educated guesses about origins in the past. Operational science has indeed been very successful in understanding the world, and has led to many improvements in the quality of life, e.g. putting men on the moon and curing diseases.

(Author's Note: The following is a letter Dr. Sarfati wrote to an enquirer who believed that atoms had to be held together by miraculous means.)

"'Natural laws' also help us make predictions about future events. In the case of the atom, the explanation of the electrons staying in their orbitals is the positive electric charge and large mass of the nucleus. This enables us to make predictions about how strongly a particular electron is held by a particular atom, for example, making the science of chemistry possible. While this is certainly an example of Col. 1:17, simply saying 'God upholds the electron' doesn't help us make predictions."

The difference between operational and origins science is important for seeing through silly assertions such as the following by Levitt (as quoted by Lerner): '... evolution is as thoroughly established as the picture of the solar system due to Copernicus, Galileo, Kepler, and Newton.'

We can observe the motion of the planets, but no one has ever observed an information-increasing change of one type of organism to another.

To explain further: the laws that govern the operation of a computer are not those that made the computer in the first place. Lerner's anti-creationist propaganda is like saying that if we concede that a computer had an intelligent designer, then we might not analyse a computer's workings in terms of natural laws of electron motion through semiconductors, and might think there are little intelligent beings pushing electrons around instead. Similarly, believing that the genetic code was originally designed does not preclude us from believing that it works entirely by the laws of chemistry involving DNA, RNA, proteins, etc. Conversely, the fact that the coding machinery works according to reproducible laws of chemistry does not prove that the laws of chemistry were sufficient to build such a system from a primordial soup.

The following definition and discussion of fact-value distinction comes from Wikipedia. Dr. Sarfati refers to the term in his discussion of Stephen Jay Gould's book.

The fact-value distinction is a concept used to distinguish between arguments which can be claimed through reason alone, and those where rationality is limited to describing a collective opinion. In another formulation, it is the distinction between what is (can be discovered by science, philosophy or reason) and what ought to be (a judgment which can be agreed upon by consensus). The terms positive and normative represent another manner of expressing this, as do the terms descriptive and prescriptive, respectively. Positive statements make the implicit claim to facts (e.g. water molecules are made up of two hydrogen atoms and one oxygen atom), whereas normative statements make a claim to values or to norms (e.g. water ought to be protected from environmental pollution).

The fact-value distinction emerged in philosophy during the Enlightenment; in particular, David Hume (1711-1776) argued that human beings are unable to ground normative arguments in positive arguments, that is, to derive 'ought' from 'is'. Hume was a skeptic, and although he was a complex and dedicated philosopher, he shared a political viewpoint with previous Enlightenment philosophers such as Thomas Hobbes (1588-1679) and John Locke (1632-1704). Specifically, Hume, at least to some extent, argued that religious and national hostilities that divided European society were based on unfounded beliefs; in effect, he argued they were not found in nature, but a creation of a particular time and place, and thus unworthy of mortal conflict. Thus Hume is often cited as being the philosopher who finally debunked the idea of nature as a standard for political existence. Alternatively, the phrase "naturalistic fallacy" is used to refer to the claim that what is natural is inherently good or right, and that what is unnatural is bad or wrong (see "Appeal to nature"). It is the converse of the moralistic fallacy, or that what is good or right is natural and inherent.

Samuel Adams believed, along with many rationalists who also acknowledged and respected biblical belief, that Natural Law originated with God at Creation, and was something He had set in motion to keep things running as they should. It included God's provisions and dealings with man, an immutable law, but one that was natural, in the sense of originating with the God of Nature. "The natural liberty of man is to be free from any superior power on Earth, and not to be under the will or legislative authority of man, but only to have the law of nature for his rule."

Alfred, Lord Tennyson and Nature Red in Tooth and Claw

The phrase "nature red in tooth and claw" comes from In Memoriam, A.H.H., a long group of poems written over many years by Alfred, Lord Tennyson completed in 1849. In it Tennyson struggled with his grief over Arthur Henry Hallam, a dear friend who was engaged to Tennyson's sister but died at age 22. The section containing the often-quoted phrase appears below. The complete work is many pages in length and can be viewed in various literature textbooks or online.

LVI
'So careful of the type?' but no.
From scarped cliff and quarried stone
She cries, 'A thousand types are gone:
I care for nothing, all shall go.'

Thou makest thine appeal to me:
I bring to life, I bring to death:
The spirit does but mean the breath:
I know no more.' And he, shall he,

Man, her last work, who seem'd so fair,
Such splendid purpose in his eyes,
Who roll'd the psalm to wintry skies,
Who built him fanes of fruitless prayer,

Who trusted God was love indeed
And love Creation's final law—
Tho' Nature, red in tooth and claw
With ravine, shriek'd against his creed—

Who loved, who suffer'd countless ills,
Who battled for the True, the Just,
Be blown about the desert dust,
Or seal'd within the iron hills?

No more? A monster then, a dream,
A discord. Dragons of the prime,

That tare each other in their slime,
Were mellow music match'd with him.
O life as futile, then, as frail!
O for thy voice to soothe and bless!
What hope of answer, or redress?
Behind the veil, behind the veil.

These poems chronicle Tennyson's struggle to understand how death fit in with the God of life. In them he also tried to deal with philosophical questions in areas including the newly-named science of Biology.

Darwin had not yet made a name for himself, but other writers were beginning to put together theories of evolution. These were based on ideas like inheritance of acquired characteristics, spontaneous generation, and vital fluids flowing through living things that forced them to undergo evolutionary changes.

All of these ideas were disturbing to thinking men like Tennyson, trying to embrace Rationalism and rely on man's reason to solve life's great questions. They also wondered how the so-called "discoveries" of randomness and chance could co-exist with the orderly Creator and loving Sustainer of the Bible. The theories listed above have all since been discredited but more have sprung up to replace them.

Tennyson's final conclusion in the same set of poems, finished in 1849, includes the following section. It is usually placed first in the published versions but was probably written last. The emphasis is added to show what Tennyson thought of his earlier doubts about how "natural law" fit in with a loving creator God. The text comes from http://www.online-literature.com/tennyson/718/).

For knowledge is of things we see;
And yet we trust it comes from thee,
A beam in darkness: let it grow.
Let knowledge grow from more to more,

But more of reverence in us dwell;
That mind and soul, according well,
May make one music as before,
But vaster.

We are fools and slight;
We mock thee when we do not fear:
But help thy foolish ones to bear;
Help thy vain worlds to bear thy light.

Forgive what seem'd my sin in me;
What seem'd my worth since I began;
For merit lives from man to man,
And not from man, O Lord, to thee.

Forgive my grief for one removed,
Thy creature, whom I found so fair.
I trust he lives in thee, and there
I find him worthier to be loved.

Forgive these wild and wandering cries,
Confusions of a wasted youth;
Forgive them where they fail in truth,
And in thy wisdom make me wise.

Solar Energy Source Appendix

We have condensed more information from two articles from the Institute for Creation Research website. The first one, "The Sun Is Shrinking," begins with a "disclaimer" of sorts which is included here in its entirety.

"The Sun Is Shrinking"

by Russell Akridge, Ph.D.

Since publication of this article in 1980, studies of the sun's size have yielded different results. Currently, scientists are not united enough concerning any broadscale trends to support age estimates based on the size of the sun. In his 1998 article "The Young Faint Sun Paradox and the Age of the Solar System," Dr. Danny Faulkner provided an updated perspective that is more consistent with the relevant solar data. Other studies do provide ample evidence for the youth of the solar system and earth, such as the studies cited in the Evidence section Many Earth Clocks Indicate Recent Creation.

It may seem at first that ICR is backing off its position on whether this article is accurate or not. More likely they are simply acknowledging that this article is somewhat dated, saying they do not currently have writers working in this particular field of study, and instead offering other young creation articles in support of the position. (We will also deal with Dr Faulkner's article in this review.) Our position is that the information in this article and Dr. Faulkner's are both accurate and present uniformitarians with real dilemmas which they are probably unwilling to address.

The following quotation appears near the beginning of the Akridge article.

"John A. Eddy (Harvard -Smithsonian Center for Astrophysics and High Altitude Observatory in Boulder) and Aram A. Boornazian (a mathematician with S. Ross and Co. in Boston) have found evidence that the sun has been contracting about 0.1% per century...corresponding to a shrinkage rate of about 5 feet per hour."

Dr. Akridge explains the significance of this statement.

"A creationist, who may believe that the world was created approximately 6 thousand years ago, has very little to worry about. The sun would have been only 6% larger at creation than it is now. However, if the rate of change of the solar radius remained constant, 100 thousand years ago the sun would be twice the size it is now. One could hardly imagine that any life could exist under such altered conditions. Yet 100 thousand years is a minute amount of time when dealing with evolutionary time scales."

In his article, Dr. Akridge presents a formula for calculating the size of the sun in the past based on this current shrinkage remaining consistent. His conclusion is that if uniformitarianism is true,

"It is amazing that all of this evolutionary development, except the last 20 million years, took place on a planet that was inside the sun. By 20 million B.C., all of evolution had occurred except the final stage, the evolution of the primate into man."

He points out that it is much more reasonable to simply say that at 100,000 BC the sun would have been twice its present size. Further calculations can be presented based on *"a balance of solar forces"* and assuming a constant shrinkage rate. In fact, figures indicate that the sun's rate of shrinkage would actually have been greater in the past, making the 20 million years figure still too far in the past.

Dr. Akridge's article presents calculations to explain whether a 2.5 feet per hour contraction of the solar surface would be enough to liberate all the solar energy measured as being present. His conclusion is that there is far more than sufficient contraction taking place to produce the observable energy output. He says that gravitational self-collapse certainly accounts for at least some of the sun's energy. According to Dr. Akridge, when uniformitarians calculate the evolution of the sun, *"all of those calculations attribute practically 100% of the sun's energy over the past 5 billion years to thermonuclear fusion."* We have read many articles where the uniformitarians dismiss the solar shrinkage argument simply by saying the sun is so big a little shrinkage doesn't matter. Others claim there are cyclic variations and any shrinkage would be offset. But, as Dr. Akridge points out,

"This (cyclic) *claim is made in spite of the evidence that the shrinkage rate of the sun has remained essentially constant over the past 100 years when very accurate measurements have been made on the size of the sun. Less accurate astronomical records spanning the past 400 years indicate the shrinkage rate has remained the same for the past 400 years."*

Dr Akridge's historical background explains that the Kelvin-Helmholtz Contraction theory was considered the most likely explanation of the sun's energy production up until the 1930's, when

"the theory of evolution began to dominate the scientific scene. Then Helmholtz's explanation was discarded because it did not provide the vast time span demanded by the theory of organic evolution on the earth. The substitute theory was introduced by Bethe in the 1930's precisely because thermonuclear fusion was the only known energy source that would last over the vast times required by evolution."

This second article is presented on the ICR website as being a more updated study than the previous one.

The Young Faint Sun Paradox and the Age of the Solar System

by Danny Faulkner, Ph.D.

Dr. Faulkner gives an illustration of the unique position of the Earth for sustaining life by comparing it with conditions on Venus and Mars, making the point that the Earth's position in relation to the sun assures it is neither too hot nor too cold for life. Dr. Faulkner rehearses the standard evolutionary theory about the origin of the solar system. He points out the commonly accepted belief that gravitational compression probably explains the early history of stars and their planets. Most uniformitarians even grant that this accounts for the sun's method of generating energy in its infancy. At some point,

"conditions in the center of the Sun permitted the conversion of hydrogen into helium through nuclear fusion. While theoretical and observational questions remain, it can be assumed for purposes of discussion, that this model approximates the truth."

Faulkner goes on to explain how this thermonuclear theory of the sun's energy production works. *"Calculation shows that it is capable of supplying the Sun's current luminosity for about ten billion years."* If, as evolutionists claim, the sun was formed 9.2 billion years ago, that puts the sun at its halfway point of life and energy use.

"This means that about half the hydrogen in the core of the Sun has been used up and replaced by helium. This change in chemical composition changes the structure of the core. The overall structure of the Sun would have to change as well, so that today, the Sun should be nearly 40% brighter than it was 4.6 billion years ago."

Evolutionists, however, agree that temperatures and luminosity would need to have been relatively consistent for life to evolve as it has. Faulkner suggests a theory, which he calls "naïve," that perhaps

"Earth began cooler than it is today and has been slowly warming with time. But this is not an option because geologists note that Earth's rock record insists that Earth's average temperature has not varied much over the past four billion years, and biologists require a nearly constant average temperature for the development and evolution of life."

Faulkner presents the theories evolutionists propose to explain how this paradox could be explained. The theories explain that early atmospheres of the planets were very different from today.

"Hydrogen was quite abundant. Much of the oxygen present would have been in the form of water. With time these atmospheres followed different evolutionary paths to become the current secondary atmospheres. The prime characteristic of the secondary atmospheres is that they are oxidized, that is, most of the hydrogen has escaped, which has forced the oxygen to recombine to form other compounds."

Explanations are offered for the differences between Mars and Venus and Earth, based on the effect their relative positions and gravity would have had on this evolution of atmosphere. When it comes to Earth, however, the story gets stranger, according to Faulkner's retelling of evolutionists' claims.

"Early life forms are supposed to have introduced free oxygen into the air and regulated the amount of other gases, such as nitrogen. As new forms of life evolved, the mix of gases in Earth's atmosphere gradually changed. Evolution proposes that the early atmosphere contained a greater amount of greenhouse gases (such as methane) than today. This would have produced average temperatures close to those today, even with a much fainter Sun. As the Sun gradually increased in luminosity, Earth's atmosphere is supposed to have evolved along with it, so that the amount of greenhouse gases have slowly decreased to compensate for the increasing solar luminosity."

Faulkner makes the very understandable statement that *"The precise tuning of this alleged co-evolution is nothing short of miraculous ... Thus the incredibly unlikely origin and evolution of life had to be accompanied by the evolution of Earth's atmosphere in concert with the Sun."* We have stated many times that Secular Humanism is a religion of mythology, so it is not surprising they demand we believe in their miraculous evolutionary processes. The problem is that they claim to be factually and scientifically correct but they produce no evidence to support this claim.

"The physical principles that cause the early faint Sun paradox are well established, so astrophysicists are confident that the effect is real." Some physicists have actually ventured theories about the biosphere being a unified, living organism capable of sustaining and adapting, though this is not a popular theory. Others try to propose a kind of symbiosis with the Sun, also unlikely. There is in some scientific circles the idea that a *"life force has directed the atmosphere's evolution through this ordeal. Most find the teleological or spiritual implications of this unpalatable, though there is a trend in this direction in physics."*

"Of course, there is a third possibility. Perhaps the Earth/Sun system is not billions of years old and so there has not been a 40% increase in solar luminosity. If Earth were recently created and designed to have the kind of atmosphere that it has now and the Sun has not changed appreciably in luminosity, then the young faint Sun paradox has been resolved. While the early faint Sun paradox does not tell us that the Solar System is only thousands of years old, it does seem to rule out the age being billions of years."

Akridge, R. 1980. The Sun Is Shrinking. *Acts & Facts*. **9 (4).**

Faulkner, D. 1998. The Young Faint Sun Paradox and the Age of the Solar System. Acts & Facts. 27 (6).

Appendix Four: Recommended Reading

Websites

Heritage Foundation *askheritage.org* "To build an America where freedom, opportunity, prosperity and civil society flourish. Public policy research organization."

Cato Institute *www.cato.org* "Increase the understanding of public policies based on the principles of limited government, free markets, individual liberty, and peace."

Discovery Institute *discovery.org*. "Explore ...technology, science and culture, reform of the law, national defense, the environment and the economy, the future of democratic institutions, transportation, religion and public life, government entitlement spending, foreign affairs."

Answers in Genesis *answersingenesis.org*. "enabling Christians to defend their faith ... answers to questions surrounding the book of Genesis, ... train others to develop a biblical worldview..."

Institute for Creation Research *icr.org*. "Scientific research from a biblical perspective ...graduate-level degree program in science education, graduate-level training in biblical education and apologetics ... Publications, Events, and Media."

Ayn Rand Center for Individual Rights *aynrandcenter.org* "advance individual rights (the rights of each person to life, liberty, property, and the pursuit of happiness) as the moral basis for a fully free, laissez-faire capitalist society."

National Rifle Association *nra.org* safety and training programs for military, law enforcement, civilian, female protection and child safety, and "a major political force ... America's foremost defender of Second Amendment rights."

The Internet Sacred Text Archive *sacred-texts.com* Every kind of public-domain material remotely connected with spiritual, philosophical, and broadly religious subjects.

Creation Research Society *creationresearch.org* Education, research, journal publication, "committed to full belief in the Biblical record of creation and early history."

World Net Daily *wnd.com* Original news articles and links to outside sources for news, opinion, commentary with a conservative emphasis.

Books
James Hannam, Web site and book *God's Philosophers: How the Medieval World Laid the Foundations of Modern Science.* Icon Books, London, 2009.

Allan Bloom, *The Closing of the American Mind.* Documentation on secular humanist influence from non-Christian but conservative perspective with extensive research and proofs.

Francis Schaeffer *The God Who Is There, The Christian Manifesto, How Should We Then Live, True Spirituality, Escape from Reason, Back to Freedom and Dignity.* Schaeffer is reformed in theology, evangelical rather than fundamentalist, left America for Switzerland. He was the first evangelical to advocate political activism opposing abortion.

Aleksandr Solzhenitsyn *Gulag Archipelago* (3 volumes) Believer imprisoned under Stalin. collected stories of other prisoners massive, well-documented work on the effects of communism on its own people.

Dr. Don DeYoung, *Thousands, Not Billions: Challenging an Icon of Evolution Questioning the Age of the Earth.* Disproves uniformitarianism by collection of individual scientific studies

George W. Dollar, *A History of Fundamentalism in America.* Church history in America from founding to early 70's emphasis on the 20th century.

David O Beale, *In Pursuit of Purity.* American church history up to 1980's emphasizing conflicts between belief and unbelief, as it affects the church.

William Evans. *The Great Doctrines of the Bible.* Brief easy to read basic Bible doctrines.

John Foxe. Foxe's Book of Martyrs. History of martyrs up to Foxe's time sections added mid-16th century.

Humphreys, D. Russell, Ph.D. *Starlight and Time, Solving the Puzzle of Distant Starlight in a Young Universe.*

Josephus, *Antiquities of the Jews.* Roman General and Jewish historian. Late first century writer from Creation to his lifetime.

Josh MacDowell, *The New Evidence that Demands a Verdict*. Conservative evangelical Christian apologist, evidence in support of the Bible's truth.

G. Campbell Morgan. Commentator Baptist preacher English, lived in America.

Michael Oard. *Frozen in Time: The Wooly Mammoth, The Ice Age and the Bible*. Recent information refuting uniformitarianism.

Antonin Scalia. Supreme Court Justice, Conservative constitutional jurisprudence.

William Warren Sweet. *The Story of Religions in America*. Well-documented, honest but liberal perspective.

J.C Whitcomb and H. M. Morris. *The Genesis Flood*. Classic scientific treatise on Earth geology.

Bibliography for Antidisestablishmentarianism

Scripture references are as follows: The Bible: The King James Version, public domain. A few verses for comparison purposes are from other translations as follows: The New International Version, from the HOLY BIBLE, NEW INTERNATIONAL VERSION Registered. NIV Registered. Copyright 1973, 1978, 1984 by International Bible Society. Used by permission of Zondervan. All rights reserved. The New American Standard Version: Scripture quotations taken from the New American Standard Bible Registered, Copyright 1960, 1962, 1963, 1968, 1971, 1972, 1973, 1975, 1977, 1995 by The Lockman Foundation Used by permission.

Antidisestablishmentarianism references hundreds of authors and works, yet one source needs special mention. The website Sacred Texts by J.B. Hare is the largest collection of public domain material of which we are aware. The entire website of over one thousand books is available for purchase on either CD ROM or DVD ROM. Most of the ancient texts used in this work are public domain books from this collection. A problem with this or any other collection is proving the validity of the primary sources. Though we do not know anything about John B. Hare, except the information posted on his website, we believe that he faithfully and accurately scanned the texts. The problem is, are the texts reliable? Since they are public domain, they are older and sometimes not the latest translations. We are confident, however, that they are acceptable. Some sources we use are books where Westerners lived among a tribe and wrote down oral traditions. Though we trust that the authors accurately recorded the oral traditions, how much 'contamination' with outside influences shaped these oral traditions? The Lore of the Whare-Wananga, a New Zealand tribe, is well documented by the translator S. Percy Smith to be older than outside influences and free of 'contamination.' Myths of the Cherokee by James Mooney, however, was published in 1900 after more than 250 years of wars and close contact with outsiders. The level of outside influence on the oral traditions of the North American Indians is impossible to measure or deny.

It should also be noted that some of the authors listed here have been accused of being pseudoarchaeologists or pseudoscientists and are largely discounted by many as scholarly sources because of the conclusions they drew from their research or the inability to substantiate some of their claims. Examples of these authors are Graham Hancock, Emmanuel Velikovsky and Thor Heyerdahl. Their conclusions are in some cases not worthy of serious consideration and some of their findings are unverifiable. However, the research they conducted and the discoveries they claim to have made, when verifiable, bear serious consideration. It is necessary to go back to verifiable evidence uncovered by archaeology, exploration and scientific discovery and to draw realistic conclusions from this evidence based on biblical understanding.

Material used from these books includes discoveries verified by repeated similar references in primary sources, documented archaeological sites which beyond question exist and testimony of ancient manuscripts accepted by scholars for hundreds of years. Some evidence cannot be substantiated because it exists in off-limits areas like the interior of China or other countries experiencing dangerous travel conditions. Presenting such claims does not attest to their truth, but in most cases these finds are part of an established pattern repeated throughout the *world.*

__________. *"1549, 1559, 1662 Acts of Uniformity." Hanover Historical Texts Projects.* History Department, Hanover College, Hanover, IN. *history.hanover.edu.*

__________. Access Research Network (*ARN.org*). (A scholarly website containing scientific research articles.)

__________. *The American Heritage® Dictionary of the English Language*, Fourth Edition. ©2000 Houghton Mifflin Company. Updated in 2003.

__________. *americanpresbyterianchurch.org*

__________. "Ancient temple found under Lake Titicaca." *BBC News.* Wednesday, 23 August, 2000, 11:04 GMT 12:04 UK.

__________. *answersingenesis.org.*

__________. *Assyrian Kings' Lists.* Various translators, various public domain texts with sources including Google Books, Wikipedia, The Internet Ancient History Sourcebook, (http://www.fordham.edu halsall/ ancient/asbook.html), and various universities which have placed public domain works online.

__________. (Atheist poster compilation) From the website *scottklarr.com.*

__________. Bethel Lutheran Church, Cupertino, CA website.

__________. *Biblefacts.org*

__________. The Book of Enoch. Translated by R.H. Charles, 1917. *The Apocrypha and Pseudepigrapha of the Old Testament*. Oxford: The Clarendon Press, 1913.

__________. *BBC online*, updated April 10, 2002.

__________. "Bible Answers." Like the Master Ministries. (Mathematical calculation from proves that 10,000 people could have been born before Adam and Eve died.) *Never Thirsty.org* website.

__________. "Boat People, a Refugee Crisis." *cbc.ca digital archives*. Broadcast May 1, 2000.

__________. "PART I THE BUNDAHIS-BAHMAN YAST, AND SHÂYAST LÂ-SHÂYAST." *Sacred Books of the East, Volume 5,* 1860. Taken from the Internet Sacred Text Archive, www.sacred-texts.com, managed by John Bruno Hare.

__________. "Cave Reveals Southwest's Abrupt Climate Swings During Ice Age." *Science Daily.com,* January 25, 2010.

__________. Church Community Services, Elkhart, IN website.

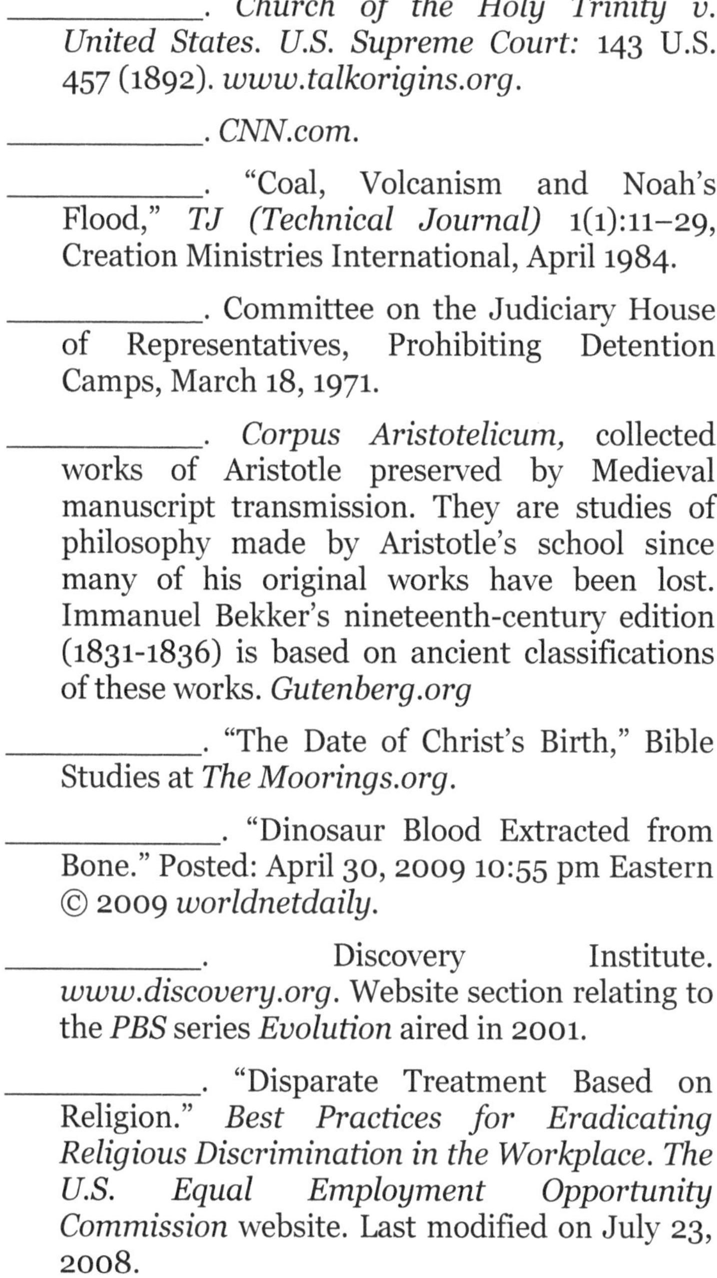

___________. *Church of the Holy Trinity v. United States. U.S. Supreme Court:* 143 U.S. 457 (1892). *www.talkorigins.org.*

___________. *CNN.com.*

___________. "Coal, Volcanism and Noah's Flood," *TJ (Technical Journal)* 1(1):11–29, Creation Ministries International, April 1984.

___________. Committee on the Judiciary House of Representatives, Prohibiting Detention Camps, March 18, 1971.

___________. *Corpus Aristotelicum,* collected works of Aristotle preserved by Medieval manuscript transmission. They are studies of philosophy made by Aristotle's school since many of his original works have been lost. Immanuel Bekker's nineteenth-century edition (1831-1836) is based on ancient classifications of these works. *Gutenberg.org*

___________. "The Date of Christ's Birth," Bible Studies at *The Moorings.org.*

___________. "Dinosaur Blood Extracted from Bone." Posted: April 30, 2009 10:55 pm Eastern © 2009 *worldnetdaily.*

___________. Discovery Institute. *www.discovery.org.* Website section relating to the *PBS* series *Evolution* aired in 2001.

___________. "Disparate Treatment Based on Religion." *Best Practices for Eradicating Religious Discrimination in the Workplace. The U.S. Equal Employment Opportunity Commission* website. Last modified on July 23, 2008.

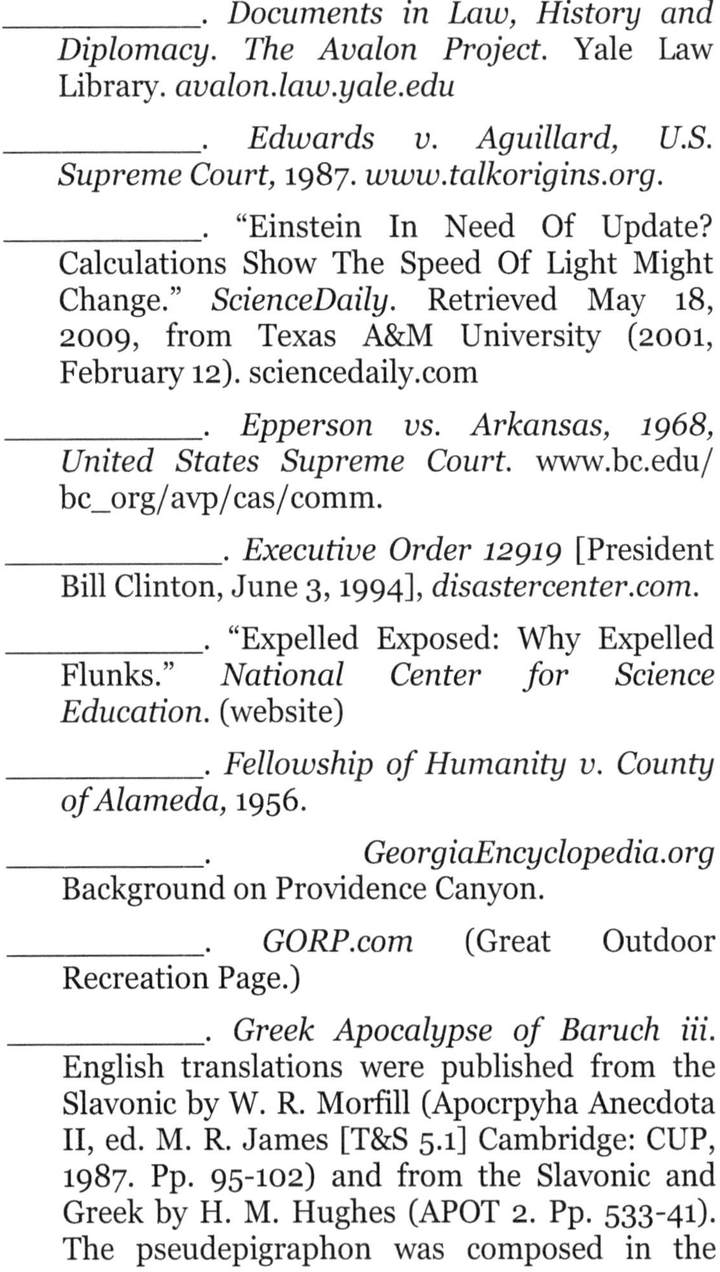

___________. *Documents in Law, History and Diplomacy. The Avalon Project.* Yale Law Library. *avalon.law.yale.edu*

___________. *Edwards v. Aguillard, U.S. Supreme Court,* 1987. *www.talkorigins.org.*

___________. "Einstein In Need Of Update? Calculations Show The Speed Of Light Might Change." *ScienceDaily.* Retrieved May 18, 2009, from Texas A&M University (2001, February 12). sciencedaily.com

___________. *Epperson vs. Arkansas, 1968, United States Supreme Court.* www.bc.edu/ bc_org/avp/cas/comm.

___________. *Executive Order 12919* [President Bill Clinton, June 3, 1994], *disastercenter.com.*

___________. "Expelled Exposed: Why Expelled Flunks." *National Center for Science Education.* (website)

___________. *Fellowship of Humanity v. County of Alameda,* 1956.

___________. *GeorgiaEncyclopedia.org* Background on Providence Canyon.

___________. *GORP.com* (Great Outdoor Recreation Page.)

___________. *Greek Apocalypse of Baruch iii.* English translations were published from the Slavonic by W. R. Morfill (Apocrpyha Anecdota II, ed. M. R. James [T&S 5.1] Cambridge: CUP, 1987. Pp. 95-102) and from the Slavonic and Greek by H. M. Hughes (APOT 2. Pp. 533-41). The pseudepigraphon was composed in the beginning of the second century A.D., but it is difficult to discover whether it was written in Greek, Hebrew, or Aramaic. (Background note from Charlesworth, James H. The Pseudepigrapha and Modern Research: with a Supplement. SBLSCS 7. Chico, Ca.: Scholars Press, 1981.)M. R. James's publication of the Greek text, until then entirely unknown, in "Texts and Studies: Contributions to Biblical and Patristic Literature," edited by J. Armitage Robinson, v., No. i., pp. 84-94, Cambridge, 1897.

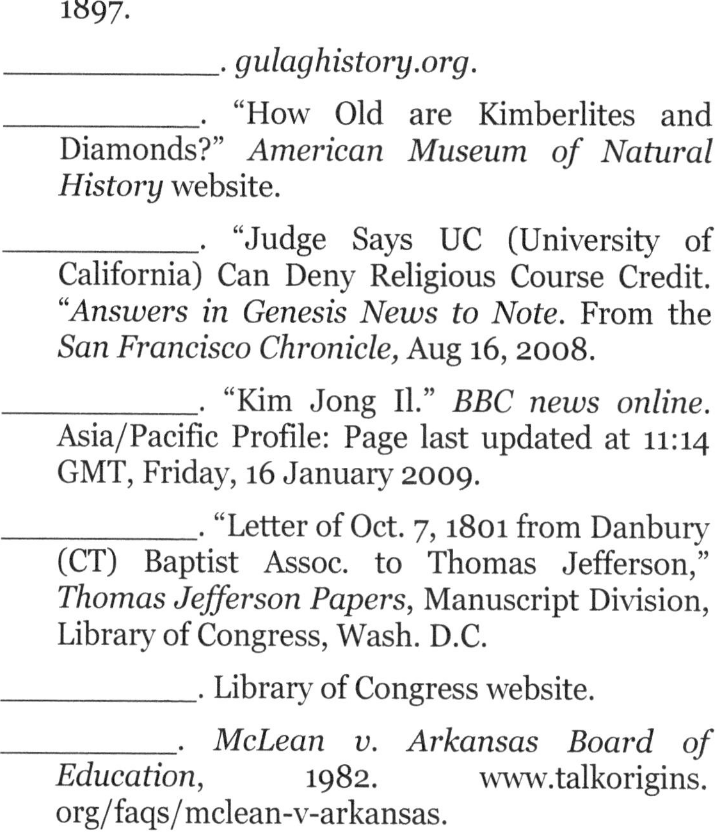

___________. *gulaghistory.org*.

___________. "How Old are Kimberlites and Diamonds?" *American Museum of Natural History* website.

___________. "Judge Says UC (University of California) Can Deny Religious Course Credit. "*Answers in Genesis News to Note*. From the *San Francisco Chronicle,* Aug 16, 2008.

___________. "Kim Jong Il." *BBC news online*. Asia/Pacific Profile: Page last updated at 11:14 GMT, Friday, 16 January 2009.

___________. "Letter of Oct. 7, 1801 from Danbury (CT) Baptist Assoc. to Thomas Jefferson," *Thomas Jefferson Papers*, Manuscript Division, Library of Congress, Wash. D.C.

___________. Library of Congress website.

___________. *McLean v. Arkansas Board of Education*, 1982. www.talkorigins.org/faqs/mclean-v-arkansas.

____________. *Magna Carta*, 1215 AD, from *The Avalon Project. Documents in Law, History and Diplomacy.* Yale Law Library. *avalon.law.yale.edu.*

____________. The Mahabharata. "Santiparva," cclx.20, 21, 23 and cxxiv.67, translated by Friedrich Max Müller and others *in Sacred Books of the East* (50 volumes), Oxford University Press, 1879-1910.

____________. *Mayflower* 1620.com (website).

____________. The National Archives. *archives.gov.*

____________. *National Geographic,* photo caption, March 1, 2010.

____________. National Park Services Website.

____________. National Park Service report on Wall Arch collapse August 4-5, 2008.

____________. *The New York Times.* News item published September 14, 1999.

____________. "NC State Paleontologist Discovers Soft Tissue in Dinosaur Bones." North Carolina State University News Release from *www.ncsu.edu/ news/press* /05-03/05 March 24, 2005.

____________. Novori.com. History and manufacture of synthetic diamonds.

____________. "Parents Fuming as Texas Schools Let Gideons Provide Bibles to Students." *Foxnews.com,* Tuesday, May 19, 2009.

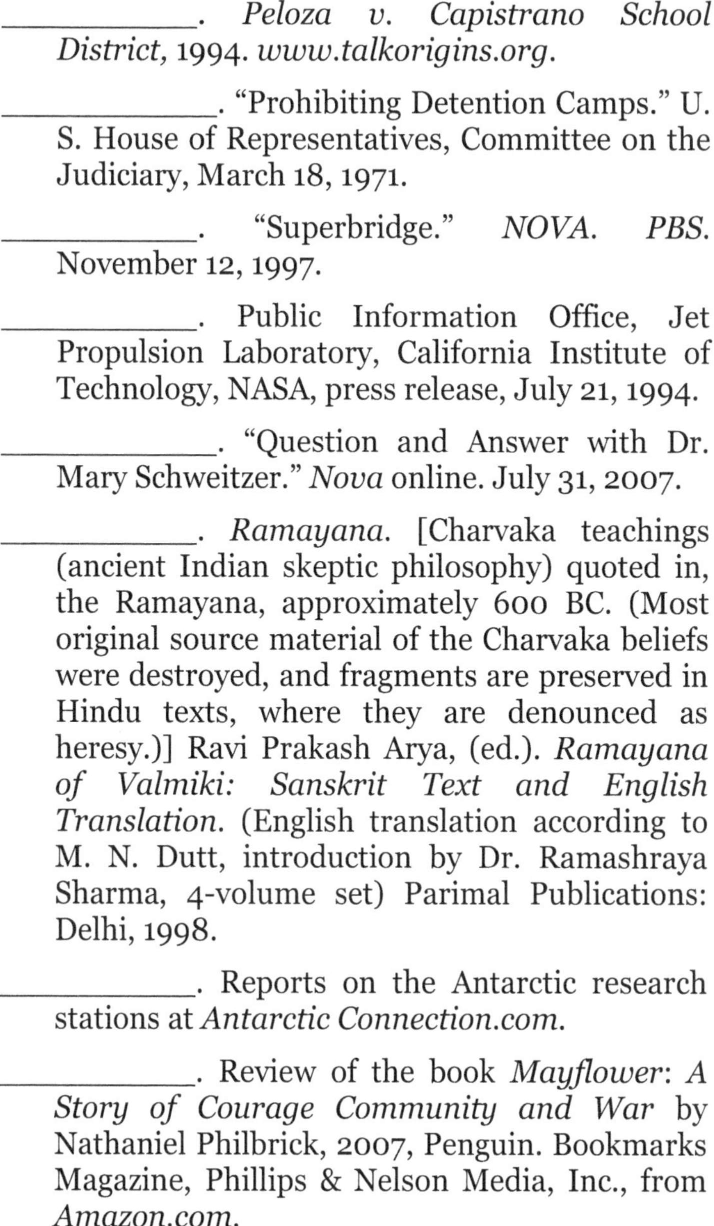

___________. *Peloza v. Capistrano School District,* 1994. *www.talkorigins.org*.

____________. "Prohibiting Detention Camps." U. S. House of Representatives, Committee on the Judiciary, March 18, 1971.

___________. "Superbridge." *NOVA. PBS.* November 12, 1997.

___________. Public Information Office, Jet Propulsion Laboratory, California Institute of Technology, NASA, press release, July 21, 1994.

____________. "Question and Answer with Dr. Mary Schweitzer." *Nova* online. July 31, 2007.

___________. *Ramayana.* [Charvaka teachings (ancient Indian skeptic philosophy) quoted in, the Ramayana, approximately 600 BC. (Most original source material of the Charvaka beliefs were destroyed, and fragments are preserved in Hindu texts, where they are denounced as heresy.)] Ravi Prakash Arya, (ed.). *Ramayana of Valmiki: Sanskrit Text and English Translation.* (English translation according to M. N. Dutt, introduction by Dr. Ramashraya Sharma, 4-volume set) Parimal Publications: Delhi, 1998.

___________. Reports on the Antarctic research stations at *Antarctic Connection.com.*

___________. Review of the book *Mayflower: A Story of Courage Community and War* by Nathaniel Philbrick, 2007, Penguin. Bookmarks Magazine, Phillips & Nelson Media, Inc., from *Amazon.com.*

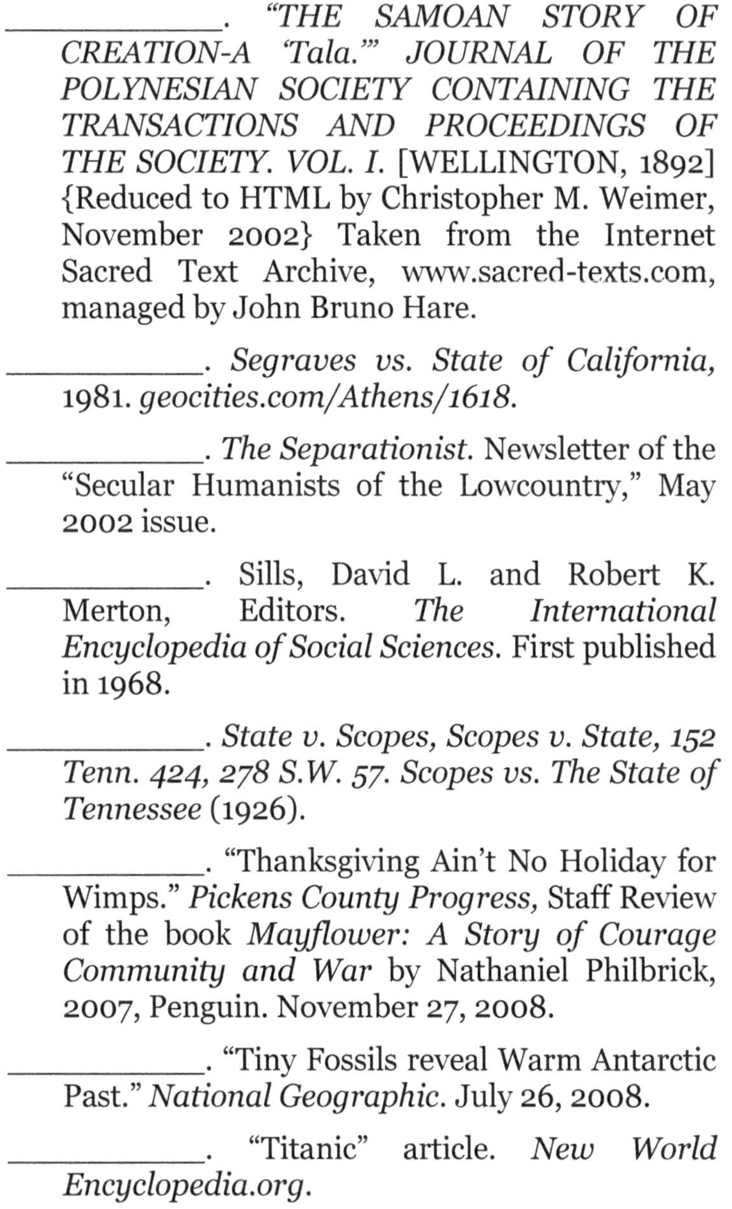

___________. *"THE SAMOAN STORY OF CREATION-A 'Tala.'" JOURNAL OF THE POLYNESIAN SOCIETY CONTAINING THE TRANSACTIONS AND PROCEEDINGS OF THE SOCIETY. VOL. I.* [WELLINGTON, 1892] {Reduced to HTML by Christopher M. Weimer, November 2002} Taken from the Internet Sacred Text Archive, www.sacred-texts.com, managed by John Bruno Hare.

___________. *Segraves vs. State of California,* 1981. *geocities.com/Athens/1618.*

___________. *The Separationist.* Newsletter of the "Secular Humanists of the Lowcountry," May 2002 issue.

___________. Sills, David L. and Robert K. Merton, Editors. *The International Encyclopedia of Social Sciences.* First published in 1968.

___________. *State v. Scopes, Scopes v. State, 152 Tenn. 424, 278 S.W. 57. Scopes vs. The State of Tennessee* (1926).

___________. "Thanksgiving Ain't No Holiday for Wimps." *Pickens County Progress,* Staff Review of the book *Mayflower: A Story of Courage Community and War* by Nathaniel Philbrick, 2007, Penguin. November 27, 2008.

___________. "Tiny Fossils reveal Warm Antarctic Past." *National Geographic.* July 26, 2008.

___________. "Titanic" article. *New World Encyclopedia.org.*

___________. *Torcaso v. Watkins. U.S. Supreme Court.*1961. *vftonline.org/* TestOath/Torcaso.htm

___________. University of Oxford, Bodleian Philosophy Faculty Library, Manuscripts and Rare Books "Medieval Manuscript Sources and Incunabula." *ox.ac.uk.*

___________. Utah Geological Survey, *Utah.gov.*

___________. *varchive.org.* A scholarly archive of Immanuel Velikovsky's unpublished works.

___________. *Voices for Evolution* (website).

___________. *Washington Ethical Society v. District of Columbia,* 249 F.2d 127 (D.C. Cir. 1957).

___________. *Web of Science* (Formerly Science Citation Index). Guillermo Gonzalez - publication record at ISU.

___________. *Webster v. New Lenox School District,* 1990, the Seventh Circuit Court of Appeals. *geocities.com/Athens/618/Webster_vs._New_ Lenox.html*

___________. *Wikipedia.org*

___________. The *Wisconsin University website,* overview of the Laramide/ Yellowstone mountain ranges with aerial maps designating geologic ages.

___________. World Net Daily. *wnd.com.*

___________. Youth Ministry Entertainment (or Y-ME ministries). *y-ment.com.*

__________. Abbott, Frank Frost and Alan Chester Johnson (authors, translators and editors). *Municipal Administration in the Roman Empire [concerning the The Law of the Twelve Tables (Duodecim Tabulae), the ancient foundation of Roman law*]. Princeton University Press, Princeton, NJ, 1926.

Adams, John. "Argument in defence of the soldiers in the Boston Massacre trial." December 1770.

__________. "Letter to Abigail Adams." July 7, 1775.

__________. "Letter to the 1st Brigade of the 3rd Division of the Militia of Massachusetts." October 11, 1798.

__________. "Letter to a friend." 1805.

__________. "Letter to Benjamin Waterhouse," 29 October 1805.

Adams, Samuel. "Letter to John Pitts." 21 January 1776.

__________. "The Report of the Committee of Correspondence to the Boston Town Meeting" Nov. 20, 1772. *history.hanover.edu/texts/adamss.html.*

Ahlstrom. Sydney F. *A Religious History of the American People.* New Haven: Yale University Press, 1972.

Aquinas, Thomas. *Summa Theologica* 1265-1274 AD Sixth Article [I-II, Q. 94, Art. 6] Objection 3. Translated by Fathers of the English Dominican Province. Benziger Brothers, New York. 1947.

Archer, Gleason Leonard, Jr. *A Survey of Old Testament Introduction.* Chicago: Moody, c. 1974. Updated and revised ed., c1994.

Arndt, William F. and F. Wilbur Gingrich, trans. Bauer, Walter. *A Greek-English Lexicon of the New Testament and Other Early Christian Literature.* University of Chicago Press: Chicago, 1967.

Asimov, Isaac. *The Roving Mind.* Prometheus Books, 1997.

Athenagoras of Athens. *Legatio pro Christianis* ["Supplication for the Christians"]. (Letter to Marcus Aurelius), 177 A.D. Translated by B. P. Pratten in "Athenagoras." *The Ante-Nicene Fathers, vol. 2,* Wm. B. Eerdmans, Grand Rapids: Michigan, 1954.

Augustine of Hippo. *City of God.* Selections. series 1, vol. 2 of the Nicene and Post-Nicene Fathers. Translated Henry Bettenson. Pelican Books, England, Clay's LTD, St. Ives Place, 1972.

Austin, Steven. Citing Hamilton Hicks. "Mineralized sodium silicate solutions for artificial petrification of wood," United States Patent Number 4,612,050, September 16, 1986, pp. 1-3. *CatastroRef*--'Catastrophe Reference Database: Catastrophes in Earth History, Geologic Evidence, Speculation and Theory', Institute for Creation Research, San Diego. Entry no. 267.

Austin, S.A. (editor). *Grand Canyon: Monument to Catastrophe.* Institute for Creation Research, Santee, California, 1994.

Ayer. A.J. (editor) *The Humanist Outlook.* Rationalist Press Association, Ltd. 1968.

Bakunin Mikhail. *God and the State.* written 1871. First published 1882 (Discovered posthumously by Carlo Cafiero and Elisée Reclus). Translated by Benjamin R. Tucker. Published by Mother Earth Publishing Association, New York, 1916.

Baldwin, James. *The Fire Next Time.* 1963 by the Dial Press. Copyright renewed 1990, 1991 by Gloria Baldwin Karefa-Smart. Published in the United States by Vintage Books, a division of Random House, first Vintage International Edition, February 1993.

Baldwin, Roger Nash. "Thirty Years Later." *Harvard Class Book of 1935.* "Baldwin's Class of 1905 on its thirtieth anniversary," Insight on the News 1997.

Balmer, Randall and John R. Fitzmier. *The Presbyterians.* Westport, CT: Praeger, 1994.

Balter, Michael. "How Human Intelligence Evolved—Is It Science or 'Paleofantasy'?" *Science* magazine, 2008.

Barnett, Randy E. "The Case for a Federalism Amendment." *Wall Street Journal,* April 23, 2009.

Bates, Mike. "Aleksandr Solzhenitsyn: The Power of One," *The National Ledger, an Eclectic Mix.* August 7, 2008. The article quotes from Aleksandr Solzhenitsyn's *The Gulag Archipelago,* 1918-1956, Volume 1, English translation by Thomas P. Whitney and Harry Willetts, Harper & Row, New York, NY, 1973.

Beale, David O. *In Pursuit of Purity*. Bob Jones University Press: Greenville, SC, 1986.

Beckford, Martin, Antony Flew, Richard Dawkins. "Flew Speaks Out: Professor Antony Flew reviews The God Delusion." (Flew's review is copyrighted as follows) Antony Flew, 2008, *bethinking.org*. *www.telegraph.co.uk/science/science-news,* 9:30 PM BST 02 Aug 2008.

Bentham, Jeremy. *The Works of Jeremy Bentham,* vol. 4, Edinburgh: William Tait. 1838-1843. 11 vols, 1843.

Bhartruhari, Neeti Shatakan (spelling varies; a work of Sanskrit philosophical verse). The empire in which he lived lasted from 185 B.C. to 135 A.D. Sahu Dharanidhar published *An English Verse Translation of Three Shatakas of Bhartruhari* in 2003.

Bierce, Ambrose. *The Enlarged Devil's Dictionary.* 1906.

Billington, Ray Allen. *Westward Expansion: A History of the American Frontier,* Macmillan, New York, NY, 1974.

Blackburn, Simon. "Independent on Sunday." *National Secular Society Newsline,* 12 May 2002.

Blackstone, William. Commentaries on the Law of England, 1765–1769.

Bonomi, Patricia U. "Religious Pluralism in the Middle Colonies." *Divining America: Religion in American History,* New York University, National Humanities Center.

nationalhumanitiescenter.org accessed May 6, 2010.

Bowden, Thomas A. "Your Child Is Not State Property," *FrontPage* Magazine, April 4, 2008. Reproduced at Ayn Rand Center for Individual Rights Website. (Thomas A. Bowden is an analyst at the Ayn Rand Institute, focusing on legal issues.)

Bradford, William. *History of Plymouth Plantation. Bradford's History of 'Plimoth Plantation' From the Original Manuscript. With a Report of the Proceedings Incident to the Return of the Manuscript to Massachusetts.* c. 1650. *Gutenberg.org*

Breasted, James Henry. *Ancient Records of Egypt: Historical Documents from the Earliest Times to the Persian Conquest,* collected, edited, and translated, with Commentary. Chicago: University of Chicago Press, 1906–1907.

Brian, Denis. Adapted from *Einstein, A Life.* John Wiley and Sons, New York, 1996.

Briggs, Charles Augustus. *American Presbyterianism.* New York, NY: Charles Scribner's Sons, 1885

Brown, Brian. *The Wisdom of the Egyptians* (1923). Taken from the Internet Sacred Text Archive, *www.sacred-texts.com,* managed by John Bruno Hare.

Budge, E. A. Wallis. "Legends of the Gods. THE HISTORY OF CREATION." *The Egyptian Texts, edited with Translation.* (Brit. Mus. Papyrus No 10,188). [1912] Taken from the

Internet Sacred Text Archive, *www.sacred-texts.com,* managed by John Bruno Hare.

Butt, Kyle, M.A. "'So We Make Up Stories' About Human Evolution." Apologetics' Press. 2008. *www.apologeticspress.org.*

Byrnes, Ryan. "Private Sector Jobs Decline, Government Jobs Increase." Quoting Bill Beach, director of the Center for Data Analysis at the Heritage Foundation. *CNS News,* Monday, March 09, 2009.

Calvin, John. Commentary on Luke 24:45. *Commentary On A Harmony of the Evangelists, Matthew, Mark, and Luke.* Translator from Latin and collator with the French version Rev. William Pringle. Edinburgh, Calvin Translation Society, 1847-1850. Calvin's Commentaries, Vol. 33: Matthew, Mark and Luke, Part III, translated by John King, 1847-50.

__________. *Institutes of the Christian Religion.* Thomas Norton, Translator. 1581.

Calabresi, Guido. *A Common Law for the Age of Statutes.* Copyright by the President and Fellows of Harvard College, 1982.

Callaway, Henry. *The Religious System of the Amazulu.* Springville, Natal, 1870.

Carson, Jonathan David. "Science's Sins of the Eyes." *New Oxford Review,* November 2001.

Castillo, Bernal Diaz Del. *The Discovery And Conquest Of Mexico* 1517-1521. Edited by Genaro Garcia, Translated with an Introduction and Notes, A. P. Maudslay. first pub 1928.

Taken from the Internet Sacred Text Archive, *www.sacred-texts.com*, managed by John Bruno Hare.

Catullus, Gaius Valerius (c. 84 – c. 54 BC). *Carmina*. Translated by Leonard C. Smithers. 1894.

Chamberlain, B.H. translator. [1882] *THE KOJIKI PART I.- THE BIRTH OF THE DEITIES. THE BEGINNING OF HEAVEN AND EARTH* Taken from the Internet Sacred Text Archive, www.sacred-texts.com, managed by John Bruno Hare.

Charron Pierre. *De la sagesse* ("Of Wisdom," In Three Parts). French version, 1601. Translated by Samson Lennard, Eliot's Court Press for Edward Blount and Will, Aspley, London, c.1615.

Chaucer, Geoffrey. "Prologue." *The Canterbury Tales*. 14th century. (Description of the Poor Parson). *msgr.ca/msgr-3/church_of_england.htm and the subsite msgr.ca/msgr-3/canterbury_tales_parson.htm.*

Clarke, Arthur C. *90th Birthday Reflections,* 2007.

__________. *Greetings, Carbon-Based Bipeds! : Collected Essays,* 1934-1998 including "Credo" (1991). St. Martin's Press, New York, NY, 1999

Cline, Aaron. Agnosticism/atheism columnist for ten years. *About.com.*

Clinton, Hillary. Speech. Global Business Coalition on HIV/AIDS Annual Awards for Business

Excellence Gala at the Kennedy Center in Washington, D.C. Wednesday, Sept. 28, 2005.

__________. Speech in San Francisco, CA. June 28th, 2004.

Coe, R.S. and M. Prevot. "Evidence suggesting extremely rapid field variation during a geomagnetic reversal." *Earth and Planetary Science Letters,* Elsevier, Amsterdam, Netherlands. Vol. 92, pp. 296-297, 1989.

Coomaraswamy, Rama. "The Conflict Between Science and Faith," from his online archives, 2001.

Covey, Stephen. *Principle-Centered Leadership.* Fireside, Simon and Schuster, Rockefeller Center, New York, NY, 1992.

Cunningham, G., Fluckiger-Hawker, E, Robson, E., and Zólyomi, G.,*The Epic of Gilgamesh, The Electronic Text Corpus of Sumerian Literature,* Oxford 1998-.

Curtis, Adrian. *Oxford Bible Atlas*, Fourth Edition. Oxford University Press: London, 2009.

Custer, Stewart. *A Treasury of New Testament Synonyms.* Bob Jones Univ. Press: Greenville, SC, 1975.

Cyprian of Carthage (3rd century AD). Letter LXXII, *Ad Jubajanum de haereticis baptizandis.* Translated by Robert Ernest Wallis. From *Ante-Nicene Fathers, Vol. 5.* Edited by Alexander Roberts, James Donaldson, and A. Cleveland Coxe. (Buffalo, NY: Christian Literature Publishing Co., 1886.)

Dalrymple, G. Brent. *The Age of the Earth*. Stanford University Press: Stanford, CA, 1991.

Darwin, Charles. *The Correspondence of Charles Darwin. 1821-1860. Vol. 8* Cambridge University Press, 1993.

__________. *The Descent of Man*. Princeton University Press, Princeton NJ, 1981.

__________. "Charles Darwin's Natural Selection," Being the Second Part of his *Big Species Book* Written from 1856 to 1858, ed. R.C. Stauffer Cambridge, 1975.

Davidson, J. P., W. E. Reed, and P. M. Davis. "The Rise and Fall of Mountain Ranges." *Exploring Earth: An Introduction to Physical Geology,* Upper Saddle River, New Jersey, Prentice Hall, 1997.

Davies, A. Powell. *America's Real Religion*. Boston: Beacon Press, 1965.

Dawkins, Richard. *The Ancestor's Tale: A Pilgrimage to the Dawn of Evolution*. (Editorial research by Yan Wong) Boston, N.Y.: A Mariner Book, Houghton Mifflin, 2004.

__________. quoted in "The Evolutionary Future of Man." *The Economist*. 1993-09-11, vol. 328.

__________. The Extended Phenotype: The Long Reach of the Gene. London: Oxford University Press, 1982, 1999.

__________. quoted in "The Flying Spaghetti Monster." Steve Paulson. *Salon.com*, October 13, 2006.

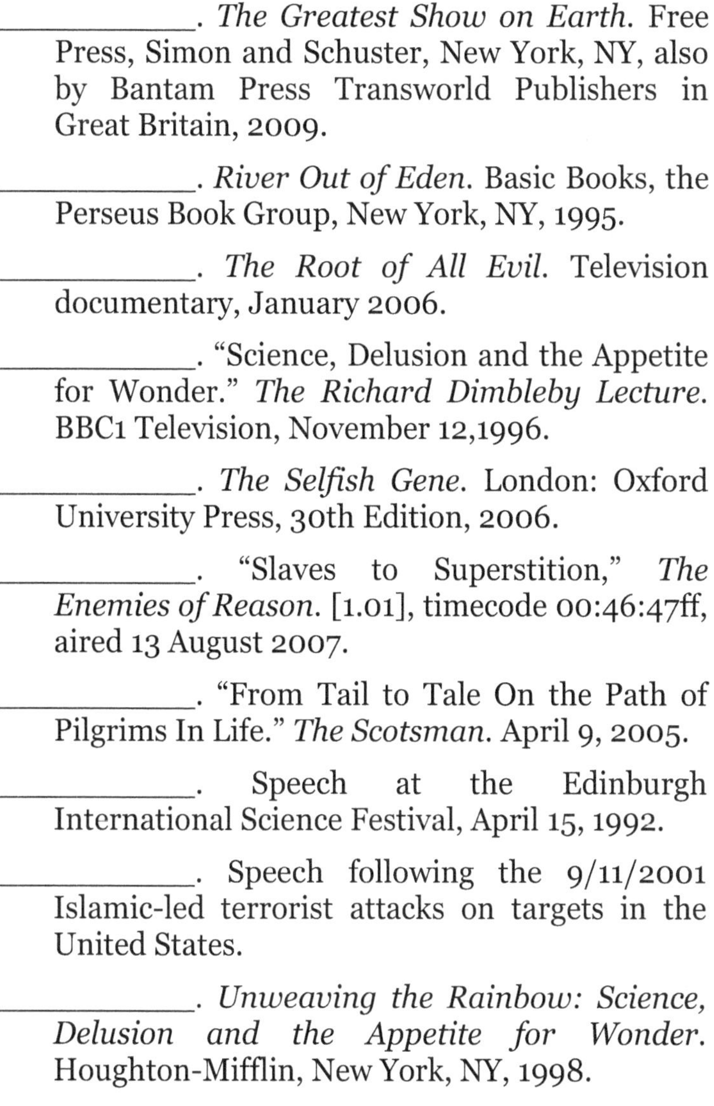

__________. *The Greatest Show on Earth.* Free Press, Simon and Schuster, New York, NY, also by Bantam Press Transworld Publishers in Great Britain, 2009.

__________. *River Out of Eden.* Basic Books, the Perseus Book Group, New York, NY, 1995.

__________. *The Root of All Evil.* Television documentary, January 2006.

__________. "Science, Delusion and the Appetite for Wonder." *The Richard Dimbleby Lecture.* BBC1 Television, November 12,1996.

__________. *The Selfish Gene.* London: Oxford University Press, 30th Edition, 2006.

__________. "Slaves to Superstition," *The Enemies of Reason.* [1.01], timecode 00:46:47ff, aired 13 August 2007.

__________. "From Tail to Tale On the Path of Pilgrims In Life." *The Scotsman.* April 9, 2005.

__________. Speech at the Edinburgh International Science Festival, April 15, 1992.

__________. Speech following the 9/11/2001 Islamic-led terrorist attacks on targets in the United States.

__________. *Unweaving the Rainbow: Science, Delusion and the Appetite for Wonder.* Houghton-Mifflin, New York, NY, 1998.

DeYoung, Dr. Don. *Thousands, Not Billions: Challenging an Icon of Evolution Questioning the Age of the Earth.* Green Forest, AR: Master Books, Inc., 2005.

Dickens, Charles. *Bleak House.* originally published serially from March 1852 to September 1853.

___________. *A Tale of Two Cities.* 1859.

Dickinson, Emily. "The Bible Is an Antique Volume," poem # 1545, Johnson, Thomas H., editor. *Complete Poems.* Boston: Little, Brown, 1960.

Dillard, Annie. *Pilgrim at Tinker Creek.* Harper's Magazine Press, New York, NY, 1974.

Disney, Walt. Quoted on *justDisney.com.*

Dollar, George W. *A History of Fundamentalism in America.* Greenville, SC: Bob Jones University Press, 1973.

Douglass, Frederick. "What, to the Slave, is the Fourth of July?" address given to a women's anti-slavery society in Rochester, New York. July 4, 1852.

DuBois, W.E.B. essay on birth control in Margaret Sanger's *Birth Control Review.* 1932.

Dunphy, John J. Quoted in *Humanist Magazine,* January-February 1983.

Dyer, B.D. and R.A. Obar. *Tracing the History of Eukaryotic Cells.* Columbia University Press, 1994.

Edison, Thomas A. "The Philosophy of Paine," a June 7, 1925 essay from the book, *The Diary and Sundry Observations,* edited by Dagobert D. Runes (1948).

Edwards, Chris. "Federal Pay Continues Rapid Ascent." *The Cato Institute Website. Cato At*

Liberty.org. The Bureau of Economic Analysis annual data on compensation levels by industry. August 24, 2009 11:57 am.

Edwards, Jonathan. "Sinners in the Hands of an Angry God." Enfield, Connecticut, July 8, 1741.

Eisenhower, Dwight David. Speech when installed as president of Columbia University in 1948.

Epicurus, from the *40 Sovran Maxims* (or "Sovereign Maxims"), 341-270 BC, as translated by Robert Drew Hicks, 1925.

___________. *Orestes*. Translated by E. P. Coleridge, 1910. Taken from the Internet Sacred Text Archive, www.sacred-texts.com, managed by John Bruno Hare.

___________. Recorded by Seneca the Younger in his *Epistle XX. From Lucius Annaeus Seneca. Moral Essays*. Translated by John W. Basore. The Loeb Classical Library. London: W. Heinemann, 1928-1935. 3 vols.

Epictetus. *The Encheiridion*. Transcribed by Flavius Arrianus. Translated by Sanderson Beck, 1911.

Eskridge, William Jr. *Dynamic Statutory Interpretation*. Copyright by the President and Fellows of Harvard College, 1994.

Eusebius of Caesarea. *Church History or Ecclesiastical History* (Hist. Ecc viii 2.) written by in the 4th century. Translated by Arthur Cushman McGiffert, From Nicene and Post-Nicene Fathers, Second Series, Vol. 1. Edited by Philip Schaff and Henry Wace. Christian Literature Publishing Co., Buffalo, NY, 1890.

Evans, William. *The Great Doctrines of the Bible.* Moody Publishers: Chicago, IL, 1995.

Fa-Hien (or Fa-Xien)). *A Record of Buddhistic Kingdoms, Being an Account by the Chinese Monk Fa-Hien of his Travels in India and Ceylon in Search of the Buddhist Books of Discipline.* Written between A.D. 399 and 412. Translated by James Legge, 1886.

Fange, Erich A. Von. "Time Upside Down." *Creation Research Quarterly*. June 1974.

Farabee, M.J. *The Online Biology Book*. Estrella Mountain Community College, Avondale, Arizona. emc.maricopa.edu.1992-2002.

Faure, G. *Principles of Isotope Geology*. 2nd. edition. John Wiley and Sons: New York, NY, 1986.

Fowler, Regi (Church/ Community Vice President). "Tolerating Thoughts On Tolerance," *Texas Sings,* Volume 13, number 2, Fall 1997.

Foxe, John. *Foxe's Book of Martyrs*. Written ca. 1560, revised in the 1700s edited by William Byron Forbush. Taken from the Internet Sacred Text Archive, www.sacred-texts.com, managed by John Bruno Hare.

Franklin, Benjamin. *Autobiography*. First English version published London, 1793. (Please see the Great Awakening Appendix for publication history.)

___________. From his speech at the Constitutional Convention Philadelphia, PA. June 28, 1787.

Freud, Sigmund. *The Future of an Illusion.* 1927. Translated by W.D Robson-Scott. English translation published by Horace Liveright and the Institute of Psychoanalysis. 1928.

Gibbs, Phil (original writer, 1996) and Sugihara Hiroshi (1997 update). "Occam's (or Ockham's) razor." University of California, Riverside, *Math.ucr.edu.*

Gibson, Rebecca. "Canyon Creation." *Answers in Genesis,* September 2000.

Gilley,Gary E. *This Little Church Went to Market–Is the Modern Church Reaching Out of Selling Out?* Evangelical Press, Carlisle, PA, July 2005.

Gish, Duane T. "A Decade of Creationist Research" (Part I). *Creation Research Society Quarterly.* 12 (1): 34-46 June, 1975.

Goetz, Delia, and Sylvanus Griswold Morley. *The Book of the People: POPOL VUH.* from Adrián Recino's translation from Quiché into Spanish. 1954. Taken from the Internet Sacred Text Archive, *www.sacred-texts.com,* managed by John Bruno Hare.

Goldberg, Justice Arthur J. The Supreme Court of the United States No. 02-1574 UNITED STATES OF AMERICA, PETITIONER v. MICHAEL A. NEWDOW, ET AL. ON PETITION FOR A WRIT OF CERTIORARI TO THE UNITED STATES COURT OF APPEALS FOR THE NINTH CIRCUIT REPLY BRIEF FOR THE UNITED STATES. June 26. 2003.

Grant, Peter R. and B. Rosemary Grant. "Genetics and the origin of bird species." *The National*

Academy of Sciences of the USA Colloquium Paper, 1997.

Green, Joey, editor. *Philosophy on the Go.* Joey Green and Alan Corcoran, Running Press, Philadelphia, PA, 2007.

Green, Nathan, Dr. *Course overview for GEO.101,"Introduction to Geology,"* University of Alabama. Spring 2006.

Grinspoon, Lester. *Marihuana Reconsidered.* Quick American Archives. Quick Trading Company, Oakland, CA, 1971.

Hall, Edward T. *Beyond Culture.* Anchor Books, Random House, New York, NY, 1976.

Hall, Fred. "Ice Cores Not All That Simple." *AEON II*: 1, 1989:199.

Haeckel, Ernst. *The History of Creation.* Vol. 1, 6-9. 1876. Translated by Joseph McCabe, Watts & Company, London, 1912.

Hamilton, Alexander. Letter to James Bayard. 1802.

Hamilton, Alexander and James Madison. *The Federalist Papers.* Signet Classics, Penguin, Putnam: New York, NY, 2003.

Hammond, James Henry (Senator of South Carolina) "Reply to Senator William H. Seward of New York." 1858.

Hancock, Graham. *Fingerprints of the Gods.* Three Rivers Press, New York, NY: Crown Publishing Group, Random House, 1995.

___________. *Underworld: The Mysterious Origins of Civilization,* Three Rivers Publishing, Crown, Random House, New York, NY, 2003.

Han Fei. c 200 BC. *The Five Vermin.* W. K. Liao (translator and annotator), The Complete Works of Han Fei Tzu. 2 vols, London, 1939-59.

Hannam, James. "Medieval Science and Philosophy" and "Science and Church in the Middle Ages." (From his Web site for the book.) *God's Philosophers: How the Medieval World Laid the Foundations of Modern Science.* Icon Books, London, 2009.

Hapgood, Charles H., J.B. Delair and E.F. Oppe. *The Path of the Pole.* Chilton Books, Philadelphia, PA, 1970.

Harris, Sam. *The End of Faith.* W.W. Norton & Company New York, NY, 2004.

Haught, James A. *2000 Years of Disbelief: Famous People with the Courage to Doubt.* Prometheus Books, Amherst, NY, 1998.

Hawking, Stephen. *The Illustrated A Brief History of Time.* New York, NY: Bantam Dell, a division of Random House, 1996.

Heinlein, Robert A. *The Notebooks of Lazarus Long,* 1978, Pomegranate Publications, Inc., 1999.

Herndon, William and Jesse W. Weik. *Herndon's Lincoln: The True Story of a Great Life,* a three volume edition published by Belford, Clarke & Company beginning in 1889.

Herodotus (484-ca. 425 BC). *Histories*. English translation G. C. Macaulay. Macmillan, London and NY, 1890.

Hesiod. *Theogeny*. Translated by Hugh G. Evelyn-White. 1914. Taken from the Internet Sacred Text Archive, www.sacred-texts.com, managed by John Bruno Hare.

Heyerdahl, Thor. *Kon-Tiki: Across the Pacific in a Raft* (The Kon-Tiki Expedition: By Raft Across the South Seas) F.H. Lyon, Translator. Rand McNally & Company: Skokie, IL, 1950.

___________. *Aku-Aku: The Secret of Easter Island*. 1958.

Hitler, Adolph. Speech given May 1, 1937.

___________. Speech given at Elbing, Germany. November 6, 1939.

Hoagland, Peter. American lawyer and congressman (US House of Representatives, Democrat, Nebraska), in a radio speech with Pastor Everett Silevan, 1983, documented in Bill Clinton: Friend or Foe? Ann Wilson, J. W. Publishing Company, 1993.

Hodges, Charles. *Systematic Theology*. 3 Volumes. Hendrickson Publishers: Peabody, MA, 1999.

Holmes, Oliver Wendell. "The Theory of Legal Interpretation." 12 Harvard Law Review. 417, 419 (1899).

Homer. *The Iliad*. C. 850 BC Translated by Samuel Butler, 1900. Gutenberg.org.

___________. *The Odyssey*.

Hornberger, Jacob G. "Your Children Are the Property of the State," *The Future of Freedom Foundation Website.* April 2000.

Howorth, H.H. *The Mammoth and the Flood: An Attempt to Confront the Theory of Uniformitarianism with the Facts of Recent Geology.* London: Sampson Low, Marston, Searle & Livingston. 1887. Reproduced by the Sourcebook Project, Glen Arm, Maryland.

Humphreys, D. Russell, Steven A. Austin, John R. Baumgardner, and Andrew A. Snelling. "Helium Diffusion Age of 6,000 Years Supports Accelerated Nuclear Decay." *Creation Research Society Quarterly Journal.* (CRSQ) Vol 41 No 1 June 2004. Creation Research.org, Copyright © 2004 by Creation Research Society.

Humphreys, D. Russell, Ph.D. *Starlight and Time, Solving the Puzzle of Distant Starlight in a Young Universe.* Green Forest, AR: Master Books, Inc., 2004, Ninth Printing.

Huxley, Aldous. "Confessions of a Professed Atheist," Report: *Perspective on the News,* Vol. 3, June 1966, p. 19.

Huxley, Julian. "At Random," a television preview on Nov. 21, 1959.

Huxley, Thomas Henry. Letter to Charles Kingsley (23 September 1860).

Ingersoll, Robert G. *Thomas Paine.* 1892. Thomas Paine National Historical Association website, http://www.thomaspaine.org/bio/ingersoll1892.html.

Iype, George. First press conference of Indian Prime Minister Manmohan Singh. *Rediff, India Abroad*, May 20, 2004.

Jay, William. "Charge to the Grand Jury of Ulster County" on Sept. 9, 1777. from *The Life of John Jay*. J. & J. Harper, New York, NY, 1833.

Jefferson, Thomas. *Autobiography*. 1821.

___________. "Draft for a Bill for Establishing Religious Freedom. Proposed to the Virginia Assembly," 1779. odur.let.rug.nl ~usa/P/tj3/writings/draft1779.htm

___________. "Letter to the Secretary of the Treasury, Albert Gallatin," 1802.

___________. *The Writings of Thomas Jefferson,* Albert E. Bergh, ed. (Washington, D. C.: The Thomas Jefferson Memorial Association of the United States, 1904), Vol. XVI, pp. 281-282.

___________. University of Virginia Library Collection of the letters and papers of Thomas Jefferson.

Jensen, Carl, *http://web.archive.org/web/20050831053419 /www.pbnnews*.reposted on the website *D.program.net* October 1, 2008,

Johnson, Allen H. "James Dale, the Supreme Court and fond memories of Troop 148," *News and Record I,* July 30, 2000. *Gay Straight Advocates for Education Website (gsafe.org).*

Johnson, Samuel. *The History of Rasselas, Prince of Abissinia*. 1759.

Josephus, Flavius. *Against Apion*. William Whiston, Translator, 1737. Taken from the Internet Sacred Text Archive, www.sacred-texts.com, managed by John Bruno Hare.

_____________. *Antiquities of the Jews.*

_____________. *Autobiography.*

_____________. *Hades.*

_____________. *Wars of the Jews.*

Justin Martyr. *First Apology.* Translated by Alexander Roberts and James Donaldson. 1867.

Keil, C. F. and F. Delitzsch. *Commentary on the Old Testament. 10 Volumes* Hendrickson Publishers: Peabody, MA, Updated Edition 1996.

King, Coretta Scott. Speech at the Palmer House Hilton in Chicago April 1, 1998.

King, Leonard William, translator. *ENUMA ELISH: THE EPIC OF CREATION (from The Seven Tablets of Creation,* London 1902) Public Domain. Taken from the Internet Sacred Text Archive, www.sacred-texts.com, managed by John Bruno Hare.

Kipling, Rudyard. *The Jungle Book.* originally published serially, 1893-1894.

Kurtz, Paul, Editor. *A Secular Humanist Declaration,* issued by The Council for Democratic and Secular Humanism (now the Council for Secular Humanism). Published in Free Inquiry Magazine, 1980.

LaBahn, Jeri. "Education and Parental Involvement in Secondary Schools: Problems, Solutions, and Effects," *Educational Psychology Interactive.* Valdosta, GA: Valdosta State University, 1995.

Landor, Walter Savage. "Melanchthon and Calvin," *Imaginary Conversations.* 1824-29.

Lee, Harper. *To Kill A Mocking-Bird.* Harper and Row, New York, NY, 1961 (copyright 1960 by the author, renewed 1988).

Lee, Robert E. (General Lee's son). *Recollections and Letters of General Robert E. Lee.* Rod and Black Publishers, St. Petersburg, Fl, 1904.

Leeuw, Nick De. Posting by contributor Monday, Nov. 10, 2008 on the *Right Michigan.com website,* email by Mount Hope Church in Lansing, Michigan attendee who witnessed the infiltration and actions of Bash Back (A Michigan-based pro-gay and lesbian organization) at the church November 9, 2008.

Lenin, Vladimir Ilyich. *Two Tactics of Social-Democracy in the Democratic Revolution.* Written June-July 1905, first published as a pamphlet in Geneva, July 1905, translated by Abraham Fineburg and Julius Katzer, published in Lenin's Collected Works, Volume 9, 1962, Moscow. Taken from Marxist Internet archive.

Lennon, John. "Imagine." Title song from the *Imagine album.* Ascot Sound Studios Tittenhurst Park and The Record Plant, New York, NY. Apple/EMI Label. 1971.

Lerner Lawrence S. *Good Science, Bad Science: Teaching Evolution in the States,* Thomas B. Fordham Foundation, Washington, DC, 2000.

Lewis, Charles. "Gay Altar Server Contests Firing Human Rights Tribunal asked to intervene." *National Post* (Canada). Tuesday, July 14, 2009.

Lewis, C.S. *The Abolition of Man or Reflections on education with special reference to the teaching of English in the upper forms of schools.* 1943. Available online at *www.columbia.edu/cu/augustine/arch/ lewis/abolition1.htm.*

Lewis, Joseph. *Ingersoll the Magnificent,* a compilation of Ingersoll's quotations, dedicated at a memorial address in 1954, published American Atheist Press, Austin TX, 1983.

Lewontin, Richard. Quoted in a review,"Billions and Billions of Demons," *The New York Review,* p. 31, January 9, 1997.

Lial, Margaret L., Charles David Miller and E. John Hornsby. *Beginning Algebra,* Harper-Collins College Division, New York, NY, 1992.

Liddell, H.G. and R. Scott, eds. *A Greek-English Lexicon.* Oxford University Press: London, 1982.

Lincoln, Abraham. "Response to Horace Greeley's abolitionist editorial." *New York Tribune,* August 22, 1862.

Linder, Douglas O. "Speech on the Occasion of the 25th Anniversary of the Scopes Trial," July 10, 2000. "State v. John Scopes" ("The Monkey

Trial")
http://www.law.umkc.edu/faculty/projects/ftrials/scopes/evolut.htm.

Lisle, Jason, Ph.D. "God and Natural Law." *Answers in Genesis,* August 28, 2006.

Livingston, Dr. David P. "Nimrod: Who Was He? Was He Godly or Evil?" *Associates for Biblical Research.* Originally published in *ABR's BIBLE AND SPADE,* 2001.

Lowder, Jeffery Jay, ed. Farrell Till et. al. "The Jury Is In: The Ruling on McDowell's 'Evidence.'" 1997-2001. *www.infidels.org.*

Lubicz, Isha Schwaller, de. *Her-Bak: The Living Face of Ancient Egypt and Her-Bak: Egyptian Initiate.* Inner Traditions, Santa Fe, New Mexico, 1978.

Lucian of Samosata. c. A.D. 125 – after A.D. 180. An Assyrian rhetorician, and satirist who wrote in the Greek language, translated by A. M. Harmon, 1936.

Lucretius (Titus Lucretius Carus). *Of The Nature of Things.* (c 95-55 BC) Translator: William Ellery Leonard, 1916. Gutenberg.org.

Lundstrom, Laurel. "Students Free to Thank Anybody Except God." *Fox News.com.* Monday, November 22, 2004,

MacAuliffe, Max Arthur (Author and translator of Sikh texts). *The Sikh Religion, DIVINE SERVICES BY GURU NANAK AND OTHER GURUS THE JAPJI, Volume 1.* Oxford University Press: London, 1909. Taken from the

Internet Sacred Text Archive, www.sacred-texts.com, managed by John Bruno Hare.

McCafferty, Phil. "Instant petrified wood?" *Popular Science.* October 1992.

MacDowell, Josh. *The New Evidence that Demands a Verdict.* Thomas Nelson: Nashville, TN, 1999.

MacRae, Andrew. *Radiometric Dating and the Geological Time Scale Circular Reasoning or Reliable Tools?* Copyright 1997-2004 [Text last updated: October 2, 1998] *Talk Origins.org.*

Madison, James. *Federalist No. 47,* quoting Montesquieu (Charles de Secondat, Baron de Montesquieu, 1689-1755), *The Spirit of the Laws, vol. 1,* trans. Thomas Nugent (London: J. Nourse, 1777).

__________. "Letter to Robert Walsh." March 2, 1819. *http://www.stephenjaygould. org/ ctrl/church-state.html.*

Malthus, Thomas Robert. *An Essay on the Principle of Population.* 1798.

Manning, Richard and Hans Beimler, writers. Directed By: Robert Wiemer. Executive Producer: Rick Berman. Created by Gene Roddenberry,"Who Watches the Watchers?" *Star Trek the Next Generation,* Season Three, Episode Four, first aired October 16, 1989.

Marcus Aurelius. *Meditations.* 167 AD. Translated by George Long. 1862.

Marx, Karl and Friedrich Engles. "Address of the Central Committee to the Communist League." London, 1850. Translated from German in the

Soviet Union, individual translators not given. Foreign Languages Publishing House, Moscow, 1951.

__________. "Contribution to the Critique of Hegel's Philosophy of Right," 1843. Published Cambridge University 1970, editor Joseph O'Malley, translators Annette Jolin and Joseph O'Malley.

Matson, Dave E. "How Good Are Those Young-Earth Arguments?" copyright 1995.on *Infidels.org*.

Maududi, Sayeed Abdul A'la. From an address given on April 13, 1939, translation on the site *IslamistWatch.org,* no translator credited.

Merrill, Eugene H. *An Historical Survey of the Old Testament*. Baker Books: Grand Rapids, MI, 1991.

Maxwell, Bill. "Intolerance as policy." *St. Petersburg Times*. August 9, 1998.

Mill, John Stuart. *Autobiography*. 1873.

Miller, Kevin and Ben Stein, writers. *Expelled: No Intelligence Allowed*. Prod. Logan Craft, Walt Ruloff and John Sullivan. Dir. Nathan Frankowski. Assoc. Prod. Mark Mathis. Ed. Simon Tondeur. © 2008 Premise Media Corporation, Rampart Films Production.

Morgan, G. Campbell. *Acts of the Apostles*. 1924.

Morton, G.R. "Young-Earth Arguments: A Second Look," 1998. *home.entouch.net*.

Montgomery, Peter. Article on *AlterNet.org*. Feb. 10, 2010.

Mooney, Chris. “Survival of the Slickest: How Anti-Evolutionists are Mutating Their Message.” *The American Prospect, Liberal Intelligence.* December 2, 2002.

Mooney, James. *MYTHS OF THE CHEROKEE. From Nineteenth Annual Report of the Bureau of American Ethnology 1897-98, Part I. COSMOGONIC MYTHS.* Taken from the Internet Sacred Text Archive, www.sacred-texts.com, managed by John Bruno Hare.

Morris, Henry M. *The Genesis Record: a Scientific and Devotional Commentary on the Book of Beginnings.* Grand Rapids, MI. Baker Book House, 1976.

Mulsow, Martin and Jan Rohls. *Socinianism And Arminianism : Antitrinitarians, Calvinists, And Cultural Exchange in Seventeenth-Century Europe,* part of the series *Brills Studies of Intellectual History,* edited by A.J. Vanderjagt, University of Gronigen, Netherlands, 2005.

Newton, Isaac. Unpublished notes for the *Preface to Opticks* (1704) quoted in *Never at Rest: A Biography of Isaac Newton* by Richard S. Westfall, Cambridge Paperback Library, 1983.

Nicholls, David. *Atheist Foundation of Australia,* undated article on the Foundation’s website.

Nietzsche, Friedrich. Human, *All-Too-Human, A Book for Free Spirits.* German version 1878. Translated by Marion Faber and Stephen Lehmann. English version published by Lincoln: University of Nebraska Press, 1984.

Oard, Michael. *Frozen in Time: The Wooly Mammoth, The Ice Age and the Bible.* Green Forest, AR: Master Books, Inc., 2004.

Ovid (Publius Ovidius Naso). *Metamorphoses.* Completed in AD 8.Translated by Henry Thomas Riley, 1851.

Paine. Thomas. *The Age of Reason,* in 3 parts, 1794, 1795, 1807. *Gutenberg.org*

___________. *Agrarian Justice,* printed in English by W. Adlard in Paris, and in London for T. Williams, No. 8 Little Turnstile, Holborn, 1797.

___________. "Answer to the Bishop of Lladaff." (Concerning The Age of Reason) published in the *Theophilanthropist,* New York, NY, 1810. [Submitted posthumously by the widow of Elihu Palmer, who attended Paine during his illness in 1806, in the house of William Carver.]

___________. *Common Sense. Gutenberg.org*

___________. "Essay on Dream." Published New York, 1807. [Full title: "An Examination of the Passages in the New Testament, quoted from the Old and called Prophecies concerning Jesus Christ. To which is prefixed an Essay on Dream, showing by what operation of the mind a Dream is produced in sleep, and applying the same to the account of Dreams in the New Testament. With an Appendix containing my private thoughts of a Future State. And Remarks on the Contradictory Doctrine in the Books of Matthew and Mark."].

___________. *Examination of the Prophecies,* pamphlet published in 1807.

___________. *First Principles of Government,* 1795.

___________. "Letter to Andrew Dean," New York, August 15, 1806.

___________. "A letter to the Hon. Thomas Erskine, on the Prosecution of Thomas Williams for publishing the Age of Reason. By Thomas Paine, Author of *Common Sense, Rights of Man,* etc. With his *discourse at the Society of the Theophilanthropists.* Paris: Printed for the Author." This pamphlet was carried through Barrois' English press in Paris, September, 1797.

___________. Memorial to James Monroe, 10 Sept. 1794.

___________. "Of The Books Of The New Testament: Address To The Believers In The Book Called The Scriptures," *Prospect Papers* Magazine (also titled "A View of the Moral World,"), 1804, published monthly by Elihu Palmer in New York.

___________. *The Rights of Man.* Gutenberg.org. 1791.

Palin, Sarah. *Going Rogue: An American Life.* Harper Collins, New York: NY, 2009.

Parsons, Thomas J., et. al. "A high observed substitution rate in the human mitochondrial DNA control region." *Nature Genetics* 15, 363 - 368 (1997).

Pasteur, Louis. Correspondence I, p. 382-383,"To the Rector of the Academia de Douai," 15 Nov. 1855. Cuny, H., *Louis Pasteur, The Man and his*

Theories, Translated P. Evans, London, The Souvenir Press, 1965.

Patten, Donald W. and Samuel R. Windsor. "Catastrophic Theory of Mountain Uplifts (A Crustal Deformation Theory)." *Catastrophism and Ancient History Vol. XIII Part 1* January 1991.

Pell, George, Cardinal. "Varieties of Intolerance: Religious and Secular," Thomas More Lecture on Religion in the Public Square, hosted by the Oxford University Newman Society, *LifesiteNews,* published March 12, 2009.

Penn, William. *THE TRYAL of WILLIAM PENN and WILLIAM MEAD*, at the Sessions held at the Old Baily in London, the 1st, 3rd, 4th, and 5th of September, 1670. *Gutenberg.org.*

Pierce, Chester M. Address at Childhood International Education Seminar, 1973.

Pitman, Sean, M.D. "Ancient Ice" (a PowerPoint presentation) created in Jan 2006. Includes testimony from a telephone interview with Bob Cardin, project manager to recover "Glacier Girl."

Plato. *The Republic.* Translated by Benjamin Jowett over a period of 30 years until his death in 1893, completed by Lewis Campbell. *Gutenberg.org.*

_________. *Critias.*

_________. *Phaedrus.*

_________. *Timmaus.*

Plutarch. *Lives.* Translated by John Dryden, 1683. Gutenberg.org.

Poincaré, Henri. “Science et méthode.” (“Science and Method”), 1908, English translation in *The Foundations of Science: Science and Hypothesis, The Value of Science, Science and Method,* The Science Press, translated by George Bruce Halstead, 1913.

Polo, Marco and Rustichello of Pisa, *The Travels of Marco Polo, Volume 1, THE COMPLETE YULE-CORDIER EDITION Including the unabridged third edition* (1903) of Henry Yule's annotated translation, as revised by Henri Cordier; together with Cordier's later volume of notes and addenda. 1920. Chapter XVII. Gutenberg.org.

Porter, Janet. “Forced Vaccines: Ready For Yours?” *Faith2action,* posted: August 18, 2009 1:00 am Eastern 2010 *World Net Daily. wnd.com.*

Prager, Dennis. “Breastfeeding as a Religion.” *World Net Daily. wnd.com.* posted November 11, 2003 1:00 am Eastern.

Protagoras of Abdera (ca. 490-ca. 420 BC) Greek philosopher, agnostic, logician, believed to be from his lost work *On the Gods*. Included in the following work: *Aristophanes. Clouds*. Intro. and trans. by Carol Poster. In Aristophanes 3, ed. David Slavitt and Palmer Bovie. Philadelphia PA: University of Pennsylvania Press, 1999.

Radest, Howard B. “Are We Religious?” By Algernon David Black. Collected in *Understanding Ethical Religion*. Produced for the American Ethical Union Library, 1975.

Rand, Ayn. *Atlas Shrugged*, author's copyright 1957. Signet, New American Library, Penguin Group, New York, NY 1996.

Randerson, James. "We Know Nothing About Brain Evolution." *Guardian* (UK). Report on a 2008 Lewontin speech titled, "Why We Know Nothing About the Evolution of Cognition."

Rantoul, Robert. Fourth-of-July address. Scituate, Massachusetts, 1836.

Reese, Lizette. 1856-1935. From the poem "Truth." *American Women Poets of the Nineteenth Century*. Anthology edited by Cheryl Walker. Rutgers University, New Jersey, 1992.

Regnerus, Mark. "Sex and the Evangelical Teen." *Forbidden Fruit: Sex & Religion in the Lives of American Teenagers*. Oxford University Press, New York, NY, 2007). THOUGHTS, "Minority report," World magazine, Vol. 22, No. 29, August 11, 2007.

Rickman, Thomas Clio. *Life of Thomas Paine,* 1819.

Robertson, A.T. *Word Pictures of the New Testament*. Broadman Press: Nashville, TN: 1932, 33, Renewal 1960.

Rolston, Bruce. "Speed Of Light May Not Be Constant, Physicist Suggests." Report on an article co-authored by University of Toronto Physics professor John Moffat and former U of T researcher Michael Clayton and published in *Physics Letters* in 1999. *ScienceDaily,* October 6, 1999.

Rooney, Andy. *Sincerely, Andy Rooney*. Essay Productions, Public Affairs, by the Perseus Group, New York, NY, 1999.

Rosenhouse, Jason. *EvolutionBlog*, Posted March 5, 2010.

Roys, Ralph L, translator. *THE BOOK OF CHILAM BALAM OF CHUMAYEL* 1933. Taken from the Internet Sacred Text Archive, www.sacred-texts.com, managed by John Bruno Hare.

Rummel, R.J. STATISTICS OF DEMOCIDE Chapter 4 "Statistics Of Cambodian Democide Estimates, Calculations, And Sources." Prepublication excerpt 1997, *Hawaii.edu.*

Rushdoony, Rousas. *The Mythology of Science.* Nutley, NJ: Craig Press, 1967.

Rushdie, Salman. A 1996 speech.

Russell, Bertrand. "Why I am Not a Christian." Lecture March 6, 1927, delivered to the National Secular Society.

Sagan, Carl. *Contact.* Pocket Books, Simon and Schuster, NY: NY, 1985.

___________. *Cosmos* television series. PBS, 1980.

___________. *Cosmos: A Personal Voyage* (Updated), television series, PBS, 1989.

___________. Interview with Charlie Rose, late-night PBS talk show host, 1996.

___________. *The Demon-Haunted World: Science as a Candle in the Dark,* Ballantine Book, Random House, New York, NY, 1996.

Sand, George (Amantine Aurore Lucile Dupin). 1804-1876 Letter to Gustave Flaubert, 14 September, 1871. Translated by A.L. MacKenzie, 1921.

Sanderson, Terry. Address as president of the National Secular Society of the UK, Dec 17, 2009.

Sanger, Margaret. *Pivot of Civilization,* 1932.

__________. *The Woman Rebel, Volume I, Number 1.* Reprinted in *Woman and the New Race.* New York: Brentanos Publishers, 1922.

Sarfati, Jonathan. "Who's Really Pushing Bad Science?" *Creation.com,* Creation Ministries International, 26 September 2000.

Saxe, John Godfrey. (1816-1887). "The Blind Men and the Elephant."

Sayers, Dorothy L. "The Other Six Deadly Sins." *Creed or Chaos,* Harcourt, Brace and Company, New York: NY, 1994.

Scalia, Antonin. *Common-Law Courts in a Civil-Law System: The Role of United States Federal Courts in Interpreting the Constitution and Laws.* THE TANNER LECTURES ON HUMAN VALUES. Delivered at Princeton University, March 8 and 9, 1995.

Schaff, Phillip. *History of the Christian Church, Volumes 5, 6 and 7.* Charles Scribner's Sons, New York, 1910.

Schaeffer, Francis A. "A Christian Manifesto." An address delivered by Dr. Schaeffer in 1982 at the Coral Ridge Presbyterian Church, Fort

Lauderdale, Florida. It is based on the book of the same title.

__________. *The God Who Is There.* InterVarsity Press: Downer's Grove, IL, 1998.

Schmid, Randolph E. (Associated Press). "Skull Suggests Interbreeding of Neanderthal and Modern Man." *The Denver Post*, January 15, 2007.

Schweitzer, Mary H. and Jennifer L. Wittmeyer, North Carolina State University; John R. Horner, Montana State University; Jan B. Toporski, Carnegie Institution of Washington Geophysical Laboratory. "Soft-Tissue Vessels and Cellular Preservation in Tyrannosaurus rex." *Science,* March 25, 2005. NC State, the N.C. Museum of Natural Sciences and the National Science Foundation funded the research.

__________. "Soft tissue and cellular preservation in vertebrate skeletal elements from the Cretaceous to the present." *Proceedings of the Royal Society of Biological Sciences.* vol. 274 no. 1607 183-197, 22 January 2007.

Schleiermacher, Friedrich. Letter to his father. January 1787. Martin Redeker, translator, *Schleiermacher: Life and Thought.* Fortress Press. 1973.

Sedgwick, Adam (Woodwardian Professor of Geology at Cambridge). "Letter to Charles Darwin." November 24, 1859.

Sellars, Roy Wood and Raymond Bragg. *Humanist Manifesto I draft*, 1933.

Semken, Steven, et. al. “Trail of Time” Exhibit, Grand Canyon National Park. Associate Professor of Geoscience Education and Geological Sciences, School of Earth and Space Exploration at Arizona State University. From the *Arizona State University website*. 2008.

Seneca, Lucius Annaeus (Seneca the Younger. 4 BC to AD 65). “A letter to Serenus,” as translated in *Tranquillity of Mind and Providence* by William Bell Langsdorf, 1900.

Serling, Rod. Last interview before his death, with Linda Brevelle, March 4, 1975.

Shakespeare, William. *Hamlet,* Act I Scene iii, Polonius to his son Laertes.

Shallit, Jeffrey. “Pamela Winnick’s Science Envy.” *Blogspot*. Monday, July 10, 2006.

Shaw, George Bernard. *Androcles and the Lion.* 1913.

Shelley, Mary Wollstonecraft. *A Vindication of the Rights of Woman With Strictures on Political and Moral Subjects*, 1792.

Shelley, Percy Bysshe. *The Necessity of Atheism* (1811), to serve as a note to the line in Queen Mab,”There is no God” (1813).

Shepherd. Jessica. “Children educated at home twice as likely to be known to social services, select committee told” and “Home pupils more likely to be known by social services and be out of work, education or training.” *guardian.co.uk*. Tuesday, 13 October 2009.

Shermer, Michael. "The Fossil Fallacy: Creationists' demand for fossils that represent 'missing links' reveals a deep misunderstanding of science." *Scientific American.* 21 February 2005.

Simon, Sidney. *Values Clarification.* Originally published 1972, Warner Books. Revised edition by Grand Central Publishing, September 1, 1995.

Simpson, George Gaylord. *The Meaning of Evolution.* Revised edition. New Haven: Yale University Press, 1967.

Smith, S. Percy, trans. *The Lore of the Whare-wananga; or Teachings of the Maori College On Religion, Cosmogony, and History.* Written down by H. T. Whatahoro from the teachings of Te Matorohanga and Nepia Pohuhu, priests of the Whare-wananga of the East Coast, New Zealand. (Smith was the F.R.G.S. President of the Polynesian Society.) Part I.-Te Kauwae-runga,Or 'Things Celestial.' New Plymouth, N.Z. Printed for the Society by Thomas Avery. -- 1913. {Reduced to HTML by Christopher M. Weimer, February 2003} Taken from the Internet Sacred Text Archive, www.sacred-texts.com, managed by John Bruno Hare.

Snelling, Andrew. "Radiocarbon in Diamonds Confirmed." *Answers in Genesis,* November 7, 2007. (This study was conducted during the RATE (Radioisotopes and the Age of The Earth) research project at the Institute for Creation Research.)

___________. "The Earth's magnetic field and the age of the Earth," first published: *Creation*

(Creation Ministries International), 13(4):44-48 September 1991.

____________. "The Recent Origin of Bass Strait Oil and Gas." *Creation,* 5 (2):43–46 March 1982.

Sparks, Muriel. *The Prime of Miss Jean Brodie.* Harper Collins, New York, NY, 1961.

Spence, Lewis. Excerpt from: *The Popol Vuh The Mythic and Heroic Sagas of the Kichés of Central America.* Published by David Nutt, at the Sign of the Phoenix, Long Acre, London [1908]. Taken from the Internet Sacred Text Archive, www.sacred-texts.com, managed by John Bruno Hare.

____________. *The Myths of Mexico and Peru.* (1913). Taken from the Internet Sacred Text Archive, www.sacred-texts.com, managed by John Bruno Hare.

Spurgeon, Charles Haddon. "Our Reply to Sundry Critics and Enquirers," *The Sword and Trowel,* Metropolitan Tabernacle, Elephant and Castle, London, Sept. 1887.

Steinem, Gloria. "Address to the Women of America," at the founding of the National Women's Political Caucus, 1971.

Sternberg, Dr. Richard. *RichardSternberg.org.*

Strong, James. *Strong's Exhaustive Concordance of the Bible.* Hendrickson Publishers: Peabody, MA, Updated Edition, 2007.

Suetonius (Gaius Suetonius Tranquillus). *The Twelve Caesars.* c. 117 138 AD. translation J. C. Rolfe, 1913-1914.

Sun Tzu, *The Art of War*. Estimated to have been written between 476-221 BC. Translated by Lionel Guiles, 1910. Gutenberg.org.

Swatt, Barbara (Preparer, Reference Intern). Adapted from,"Themis, Goddess of Justice," *Marian Gould Gallagher Law Library, University of Washington School of Law*. Updated Oct. 31, 2007.

Sweet, William Warren. *The Story of Religions in America*. Harper & Brothers, New York, N.Y., 1930.

Sykes, Bryan. *The Seven Daughters of Eve: The Science That Reveals Our Genetic Ancestry*. W.W. Norton: New York, N.Y., 2001.

Syrett, Harold C. editor. *The Papers of Alexander Hamilton*. NY: Columbia University Press, 1979. Vol. XXI, pp. 402-404.

Tacitus. *The Annals of Imperial Rome*. 109 AD, XIII. 32. Translated by Alfred John Church and William Jackson Brodribb, 1876.

___________. *The Histories*, 109 AD. Translated by Alfred John Church and William Jackson Brodribb, 1876.

Tamny, John. "Where Are the Supply-Side Democrats?" *National Review Online*. November 18, 2005.

Taylor, Paul (Series Editor) and Elizabeth Deane (Program Executive Producer). *American Experience, Ulysses S. Grant. PBS,* WGBH Educational Foundation, 2002.

Alfred, Lord Tennyson. *In Memoriam, AAH.* 1849. http://www.online-literature.com tennyson/718/.

Tenzin Gyatzo, 14th Dalai Lama, leader of Tibetan Buddhism,"Compassion and the Individual: the Purpose of Life." From the *Dalai Lama website.*

Thapar, Prof. Romila. *Frontline* magazine. Volume 18 - Issue 19, Sep. 15 - 28, 2001.

Thayer, Joseph Henry. *Thayer's Greek-English Lexicon of the New Testament.* Zondervan: Grand Rapids, MI, 1970.

Thiele, Edwin. *The Mysterious Numbers of the Hebrew Kings.* Zondervan Publishing House, Grand Rapids, MI, 1983.

Thiessen, Henry Clarence. *Lectures in Systematic Theology.* Wm. B. Eerdmans: Grand Rapids, MI, Revised ed., 2006.

Traufetter, Gerald. "Europe's 'Human Zoos' -- Remains of Indigenous Abductees Back Home after 130 Years." *International: Zeitgeist. Archive Der Spiegel,* 1/13/2010.

Trudeau, G.B. *Doonesbury.* Strip published in November, 1995. Universal Press Syndicate. *Doonesbury* was launched October 26, 1970.

Unruh, Bob. "Homeschooler flees state custody: Melissa Busekros surprises parents at 3 a.m." Posted: April 23, 2007 12:33 pm Eastern. *World Net Daily.*

__________. "Homeschoolers on run win U.S. asylum Judge: Teaching children 'basic right no country has right to violate.'" January 26, 2010 11:02 pm Eastern. *World Net Daily*.

Ussher, James. *Annales veteris testamenti, a prima mundi origine deducti* ("Annals of the Old Testament, deduced from the first origins of the world"), 1650.

Ustinov, Peter. Interview with Mike Wallace. March 29, 1958.

Virgil. *The Aeneid*. c. 29 BC. Translated by John Dryden 1697. *Gutenberg.org*.

__________. *Georgics,* Book Two, published c. 29 BC. Poetic translation by John Dryden, 1697. Gutenberg.org

Vega, Garcilaso de la ("El Inca," real name Gómez Suárez de Figueroa). *Comentarios Reales de los Incas*. Lisbon, 1609. Translated by Harold V. Livermore. 1965.

Velikovsky, Emmanuel. *Ages in Chaos*. Doubleday, New York: New York, 1952.

Voltaire (François-Marie Arouet). "Of Modern Atheists, Reasons of the Worshipers of God." *Atheism I, Section I. C.* 1764. Selected and Translated by H.I. Woolf, Knopf, New York, NY, 1924.

Wald, George. "The Origin of Life," *Scientific American,* 191:48, May 1954.

Walvoord, John and Roy B. Zuck. *The Bible Knowledge Commentary, Old and New Testaments*. Cook Communications: Colorodo Springs, CO, 1989.

Watts, Charles. "The Secularist's Catechism." complied in an undated book published by Watts & Co. entitled: *Pamphlets by Charles Watts Vol. I.* 1896.

Weinberg, Steve. "A Designer Universe?" *Address at the Conference on Cosmic Design, American Association for the Advancement of Science,* Washington, D.C. April 1999.

West, E. W., translator. 1880. *PAHLAVI TEXTS.* (Persian language works from c. 180 to 880 AD) Taken from the Internet Sacred Text Archive, *www.sacred-texts.com,* managed by John Bruno Hare.

Willebrands, Johannes Cardinal. *Response to the Boy Scouts of America official position on the admission of homosexual members and leaders,* 2000.

Willette (Site User). *Askville.Amazon.com.* posted late 2009 or early 2010.

Williams, Roger. "Mr. Cotton's Letter Lately Printed, Examined and Answered," (1644), and "The Hireling Ministry, None of Christ's." *The Complete Writings of Roger Williams.* The Narragansett Club (1652).

Wilson, Edward Osborne. *On Human Nature.* Harvard University Press, 1979.

Winthrop, John. From "A Model of Christian Charity," 1630. *The Avalon Project. Documents in Law, History and Diplomacy.* Yale Law Library. avalon.law.yale.edu.

Whedon, Joseph Hill "Joss." Commentary on *Buffy the Vampire Slayer* series DVD, episode 5.16 (

"The Body") (Season 5, released December 9, 2003), and an interview by Tasha Robinson for *The Onion,* (an online satirical magazine) September 5, 2001.

Wheeler, Charles N. Interview with Henry Ford. *Chicago Tribune*, May 25, 1916.

Whitcomb, J.C. and H. M. Morris. *The Genesis Flood.* Grand Rapids, MI: Baker Book House, 1961.

Whitman, Walt. "Song of Myself." From *Leaves of Grass*, first published 1855. Revised and republished many times until the "deathbed" edition finished in 1892. "Definitive version" published in 1900.

Wysong, Pippa. "Dinosaur Remains Yield Soft Tissue." *Access Excellence.* Raleigh, NC April 29, 2005. (Access Excellence is an online publication of The National Health Museum, Atlanta GA.)

Xenophanes, pre-Socratic philosopher (570-475 BC). Diels, Hermann. *Die Fragmente der Vorsokratiker* ("Pre-Socratic Fragments"). Translated by Rev. Walther Kranz. Berlin: Weidmann, 1972-1973.

Xenophon. "On Hunting." (430-354 BC). *Xenophon in Seven Volumes.* 7. Translated by E. C. Marchant, G. W. Bowersock. Constitution of the Athenians. Harvard University Press, Cambridge, MA; William Heinemann, Ltd., London. 1925.

Zahn, Drew. "Pastor waits for final word in Bible study citation: Couple ordered to get permit to host friends not out of woods yet." *WorldNetDaily* Posted: June 01, 2009 10:07 pm Eastern.

__________. "State moves to restrict Catholics in politics. Official contends church must register as 'lobbyist' to speak out." *Faith Under Fire,* Posted: June 01, 2009 9:30 pm Eastern, *World Net Daily.*

The best gift you can give an author

is an honest, thoughtful review. Please consider leaving one online. Help us understand what you liked and didn't like about the book and why. Help authors reach more readers and spread your influence and ours. If you liked the book, please recommend it to your spouse, friends, pastors, teachers, cashiers, employers, – anybody and everybody you see each day. If you don't know what to say, remember Proverb 16:3 – Commit thy works unto the Lord and thy thoughts shall be established. Thank you!

OTHER BOOKS AND PRODUCTS FROM FINDLEY FAMILY VIDEO PUBLICATIONS

All our books (including Historical Fiction, SciFi, contemporary relationships short stories, and an Archaeological Mystery serial) are linked on our blog.

Elk Jerky for the Soul includes posts on current issues, excerpts from our fiction and nonfiction works, Bible teaching, travel and everyday observations, and more.

http://findleyfamilyvideopublications.com/

Visit our YouTube Channel

https://www.youtube.com/channel/UCGhwNpU115ARMwgYwTIJBrA/featured. Book trailers, video excerpts, project teasers, and more. Science, History, Literature, and biblical worldview studies are the focus of our book and video projects.

Historical Fiction

by Michael J. Findley

The Ephron the Hittite Series (Including boxed set of all titles)

Ephron Son of Zohar

Tawananna Daughter of Zohar

Heth Son of Canaan Son of Ham, Noah

Shelometh Daughter of Yovov Wife of Ephron

Zita Son of Ephron and Shelometh

Adult Romantic Suspense

by Mary C. Findley

The Men of the Realmlands series

Book One: The Baron's Ring

Book Two: The Captain's Blade

Send a White Rose

Chasing the Texas Wind

Carrie's Hired Hand (novella)

Young Adult Historical Adventure

by Mary C. Findley

Hope and the Knight of the Black Lion (plus illustrated version)

The Benny and the Bank Robber Series

Benny and the Bank Robber (Plus homeschool editions for student and teacher with review and vocabulary)

Doctor Dad

The Oregon Sentinel

Lines in Pleasant Places

Science Fiction and Fantasy

by Michael J. Findley

The Empire Saga (all six of the following books in one volume)

City on a Hill and Sojourner (Combined Novella and Short Story)

Nehemiah LLC (Full-length novel available as a standalone ebook, paperback, and hardcover versions)

Empire One: Humiliation

Empire Two: Repentance

Empire Three: Sanctification

Steampunk

by Sophronia Belle Lyon (pen name for Mary C. Findley)

The Alexander Legacy Steampunk Literary Tribute Series

Book One: A Dodge, a Twist, and a Tobacconist (including illustrated version)

Book Two: The Pinocchio Factor

Book Three: The Most Dangerous Game

Book Four: Beware the Bustle

Fantasy/Allegory

by Mary C. Findley

Allegorical clockwork novella inspired by Little Red Riding Hood

The Acolyte's Education

A Paranormal Urban Fantasy serial

His Sign: The Wait Is Over

His Sign 2: The Ezra Solution

Contemporary Fiction

by Mary C. Findley

Romantic Suspense Novella

Fall On Your Knees

Relationships Short Stories

Fifty Shades of Faithful

Fifty Shades of Faithful 2: In Living Color

The Great Thirst Serial Archaeological Mystery (including boxed set of all titles)

Part One: Prepared

Part Two: Purified

Part Three: Pursued

Part Four: Persecuted

Part Five: Persevering

Part Six: Protected

Part Seven: Prevailing

Murder Mystery

Mapped Out Murders

Nonfiction

by Mary C. Findley

Write for the King of Glory, 2nd Edition (updated, with tips on indie writing and publishing)

by Michael J. and Mary C. Findley

The Good, the Bad, and the Ugly: A Readers' and Writers' Guide for Believers

Biblical Studies (Teacher and student editions plus excerpts in OT and NT Manuscript History)

Antidisestablishmentarianism (illustrated and plain versions)

Serial versions, illustrated and plain

What Is an Establishment of Religion?

What Is Secular Humanism?

What Is Science?

What Are the Results of the Establishment of Secular Humanism?

The Conflict of the Ages series (All have teacher and student editions plus one combined teacher edition for 1-3)

I. The Scientific History of Origins

II. The Origin of Evil in the World that Was

III. They Deliberately Forgot: The Flood and the Ice Age

IV. Ice Age Civilizations

V. The Ancient World

by Michael J. Findley

Short Recaps of longer nonfiction works (Antidisestablishmentarianism and Conflict of the Ages)

Disestablish: An Overview from Creation to the Ice Age

Under the Sun: The Truth about History from the Beginning

*****

Christian Books in Multiple Genres. Join Christian Indie Author ~ Readers Group on Facebook.

https://www.facebook.com/groups/291215317668431/

www.ingramcontent.com/pod-product-compliance
Lightning Source LLC
LaVergne TN
LVHW050525160826
845677LV00011B/1960
9798230383147